CRITICAL STUDY OF AGRICULTURAL PRODUCTIVITY IN UTTAR PRADESH

In the memory of my younger brother Brij Nath Pandit who died at the age of eighteen years on 19 July 1963 at Pahalgam (Kashmir) due to a cloud burst

CRITICAL STUDY
OF
AGRICULTURAL PRODUCTIVITY
IN
UTTAR PRADESH
1951—1975

SOM NATH PANDIT

CONCEPT PUBLISHING COMPANY
NEW DELHI

First Published 1983

Som Nath **Pandit** (*b*. 1942-)

Published by

Naurang Rai
Concept Publishing Company
H-13, Bali Nagar
NEW DELHI-110015 (India)

Printed by

Bimal Printers,
Padam Nagar
DELHI-110007

Preface

AGRICULTURAL development occupies an important place in the economic development of Uttar Pradesh. In this book an attempt has been made through analytical and graphical description to trace the broad trends in area, production and productivity of principal crops grown in Uttar Pradesh on the basis of time series data. Besides, quantifying price responsiveness rough estimates of future availability of food grains also have been attempted. Planners of agricultural development in Uttar Pradesh will find the analysis useful in particular and research workers in general.

My interest in the subject aroused when I was a student of University of Delhi and had the opportunity to listen to a series of lectures delivered by Professor A.M. Khusroo.

With humble submission I express my gratitude to Dr. S.K. Goyal, Head of Department of Economics, J.V. Jain College, Saharanpur; Dr. M.T.R. Sarma, Advisor, Labour and Employment, ILO, Bangkok; Dr. Prem Shankar Vashist, Senior Economist, NCAER, New Delhi; Dr. G.R. Saini, Economist, Ministry of Agriculture and Irrigation, New Delhi; Dr. V.N. Pandit, Delhi School of Economics; Dr. K. Subbarao, Dr. S.K. Ray and Dr. P.C. Verma, Institute of Economic Growth, Delhi; and Shri Chandresh Kumar Shrivastava, Shyam Lal College, Delhi, for their generous help in preparation of the book.

I thank Dr. Dayal Prasad Octania, Director General, State Planning Institute, Lucknow for his help in arranging the data to fill the missing links. Mr. Rajinder Kumar Agrawal an old student of mine offered his assistance at various stages of preparation of the book. I am pleased to acknowledge with thanks the co-operation of the staff of the following libraries in Delhi and New Delhi :

Directorate of Economics and Statistics, Indian Council of Agricultural Research, Planning Commission, Central Statistical Organisation, Agricultural Economics Centre, Delhi School of Economics, Jawaharlal Nehru University, Fertilizer Association of India and National Council of Applied Economic Research.

I thank Concept Publishing Company, New Delhi for publishing my book well in time.

My wife Smt. Rajni Pandita (M.A. Economics ; B.Ed.) was a source of encouragement in moments of frustration.

Needless to say, I alone am responsible for any errors which might have remained in the book.

SOM NATH PANDIT

Post-Graduate
Department of Economics
D.A.V. College
BULANDSHAHR

14 November, 1982

Contents

PREFACE v

LIST OF TABLES viii

1. INTRODUCTION 1

2. STUDY OF TRENDS IN AREA, PRODUCTION AND YIELD OF SELECTED CROPS 17

3. CHANGES IN CROPPING PATTERN AND LAND UTILISATION 54

4. STRUCTURE AND BEHAVIOUR OF PRICES OF SELECTED AGRICULTURAL COMMODITIES 78

5. SUPPLY RESPONSE OF RICE, WHEAT AND SUGARCANE 103

6. PROJECTIONS OF FOOD AVAILABILITY 127

7. SUMMARY AND CONCLUSIONS 134

APPENDICES 137

BIBLIOGRAPHY 147

INDEX 152

List of Tables

1.1. Literacy in Uttar Pradesh, 1971 *13*

2.1. Annual area variation of selected agricultural crops in Uttar Pradesh : 1951-52 to 1977-78 *22*

2.2. Index numbers of area of selected agricultural crops in Uttar Pradesh 1950-51 to 1976-77 *29*

2.3. Percentage change in area in 1976-77 over 1950-51 *31*

2.4. Annual average trend rates of production judged by two-three year averages, 1950-51 to 1952-53 to 1974-75 to 1976-77 of selected crops *32*

2.5. Annual rates of growth of output of selected agricultural commodities in Uttar Pradesh : 1952-53 to 1976-77 *34*

2.6. Three yearly moving averages of volume of output of selected crops *36*

2.7. Index numbers of production of selected agricultural crops in Uttar Pradesh, 1950-51 to 1976-77 *40*

2.8. Three yearly moving averages of yields of selected agricultural crops in Uttar Pradesh, 1951-52 to 1976-77 *42*

2.9. Index numbers of yield of selected agricultural crops in Uttar Pradesh, 1950-51 to 1976-77 *44*

2.10. Variability of yields of selected crops, 1950-51 to 1976-77 *51*

2.11. Growth rates of area, production and yield of various groups in U.P., 1950-51 to 1976-77 *52*

3.1. Cropping pattern in Uttar Pradesh, 1950-51 to 1974-75 *57*

3.2. Land use pattern in Uttar Pradesh, 1950-51 to 1975-76 *63*

3.3. Per capita sown area in Uttar Pradesh *65*
3.4. Sowing and harvesting seasons of selected agricultural crops in Uttar Pradesh *66*
3.5. Distribution of monthwise rainfall in Uttar Pradesh, 1950-51 to 1976-77 *68*
3.6. Yearwise dispersion of rainfall in Uttar Pradesh, 1950-51 to 1976-77 *70*
3.7. Irrigated area by source of irrigation *72*
3.8. Irrigated area under selected crops in Uttar Pradesh, 1950-51 to 1975-76 *73*
3.9. Distribution of fertilizer in Uttar Pradesh, 1950-51 to 1976-77 *75*
3.10. Area under HYVP crops in Uttar Pradesh, 1968-69 to 1976-77 *76*
4.1. Price indices of farm (harvest) price of paddy, wheat and sugarcane *84*
4.2. Price indices of farm (harvest) price of barley, gram and maize *89*
4.3. Farm harvest prices of selected crops in 1955-56 and 1975-76 *95*
4.4. Wholesale price indices of selected crops, 1951-52 to 1975-76 *97*
5.1. Acreage response equations of rice *118-19*
5.2. Acreage response equations of wheat *120-21*
5.3. Acreage response equations of sugarcane *122-23*
5.4. Estimated coefficient of adjustment *124*
6.1. Growth of population in Uttar Pradesh and India : 1921-71 *128*
6.2. Production of food grains in Uttar Pradesh, 1950-51 to 1975-76 *129*
6.3. Net availability of food-grains in Uttar Pradesh : 1950-51 to 1975 76 *130*
6.4. Ratio of per capita net availability of cereals, pulses and total food-grains in Uttar Pradesh to India *131*
6.5. Projected food availability : 1980-81 to 1990-91 *132*
6.6. Projected net availability of cereals, pulses and total food-grains based 1980-81 to 1990-91 *133*

1

Introduction

TODAY the countries of the world are in a race for power and economic supremacy. A frantic desire to attain the same stage of economic development as reached by the developed countries has dawned upon the underdeveloped* countries too. These underdeveloped countries have been characterised as those having a lower per capita income than the economically advanced countries like the United States of America, the United Kingdom, the Union of Soviet Socialist Republic, France and Japan. India which is rated as an underdeveloped country has also launched the programmes of economic development through its Five Year Plans with greater emphasis on agricultural development. Our country is remarkably rich in potentials for economic development and is endowed with vast reserves of natural resources. The proper utilisation of these potentials has become the chief goal of the present economic development programmes.

In our country the agricultural sector can be a major source of economic growth. In his presidential address at the All India Agricultural Economics Conference in Baroda in December 1959, Dr. S.R. Sen had pointed out correctly that in our country agriculture can help the development of our economy from the stagnant to the progressive stage in three ways :

(*i*) by increasing the gross national product,

(*ii*) by supplying the physical surplus in the shape of food and raw materials, and

---

* Most of the modern economists use the term 'developing' instead of 'underdeveloped'.

(*iii*) by providing the economic surplus which constitutes the material basis for economic development.

Dr. Sen emphasized that from the point of view of economic development, increase in production is not enough. It has to be accompanied by an increase of surplus and thereby of investment at a compound rate which should be well above the rate of increase in population.

The planners of the country while preparing Five Year Plans and Annual Plans have kept the above facts in view by according top priority to agriculture. It was only in the Second Five Year Plan that the industry took the place of agriculture in terms of priority. In the First Five Year Plan it is clearly mentioned, "Without a substantial increase in the production of food and raw materials for industry, it would be impossible to sustain a higher tempo of industrial development. In an underdeveloped economy with low yields in agriculture, there is no real conflict between agricultural and industrial development. One cannot go far without the other, the two are complementary. It is necessary, however, on economic and other grounds, first of all to strengthen the economy at the base and to create conditions of sufficiency and even plentitude in respect of food and raw materials".[1] In the Second Five Year Plan, agricultural programmes were intended to provide adequate food to support the increased population and the raw materials needed for a growing industrial economy and also to make available larger exportable surpluses of agricultural commodities.[2] In the scheme of development during the Third Five Year Plan the first priority necessarily belongs to agriculture. Experience in the first two Plans, and especially in the Second, has shown that the rate of growth in agricultural production is one of the main limiting factors in the progress of national economy.[3] In the subsequent plans the planners continued according top priority to agriculture as can be seen from the reports of Five Year Plans and Annual Plans.

It is beyond doubt that special importance is being given to agricultural development in India. In our country the scope for extending the cultivable area is limited, the emphasis is to be laid on increased production through proper investment. During the decade 1950-51 to 1960-61 the increase in agricultural production was achieved through the extension of cultivable area, but soon after it the emphasis shifted to increased productivity. Development of agriculture is the logical and necessary starting point for the general economic development of our country. Our country is predominantly agricultural and overpopulated and suffers from an appalling inequality of incomes between its rural and urban sectors. Reduction of this disparity of income demands that high priority may be accorded to agriculture. In our country the rural poor can benefit from increased productivity programmes relating to agriculture. United Nations expressed the same views about agricultural development in underdeveloped countries when it observed : "Our rapid and unbalanced growth of the industrial sector, unaccompanied by complementary changes in the agricultural sector, may give rise to phenomena which in the long run is likely to retard economic development, balance of payments' difficulties, inflation, excessive urbanisation, the disruption of social patterns.[4] While making a case study of India, Coale and Hoover came to the conclusion that "very substantial progress in that most backward part of the economy (agriculture) is a pre-requisite to successful development of the economy as a whole and that if one sector limits the growth of the other, it is more likely to be a case of agricultural growth limiting non-agricultural than vice versa.[5]

Agricultural sector, besides being a prime source of food, is also a source of raw materials for the expanding industries. Agricultural development would lead to an increase in the purchasing power of the rural poor and will help the growth of non-agricultural sector by providing a market for increased production of industries. Agricultural sector carries the double obligation ; to increase production, and to provide capital for other sectors in order to promote economic growth, while at the same time, it must provide for the welfare of the farmers and their families. The farm sector also provides for other sectors of the economy. Thus it is agriculture which must provide greater employment, either within itself or by providing capital to non-farm jobs. The capacity of absorption of increased labour force is limited and therefore there is a need for rapid expansion of non-agricultural sector in our country. Expansion of non-farm sectors requires huge amounts of capital. Additional capital has to come from the farm sector and it is possible through increased agricultural productivity. Agricultural development has to come first if the economy is to cease being a subsistence economy and to become a market economy. If the productivity in the farm sector goes sufficiently high, the nation may enjoy a food surplus of such a magnitude as to permit increased exports on favourable terms with corresponding favourable effect on the balance of payments and domestic industrialisation.

Increased agricultural productivity is the pivot on which the development of the other sectors revolves. Increased agricultural output and productivity will tend to contribute substantially to an overall economic growth of our country. In the initial stages productivity in the farm sector can be increased even with lesser amount of capital investment.

The importance of agriculture in the context of economic development has been debated upon since long. The physiocrats extolled agriculture as the only part of the economy that produced a surplus above the current requirements of labour and capital employed. The fundamental physiocratic proposition was that the farm sector alone produced an economic surplus or net product over the costs of production and, therefore, it played most strategic role in a nation's economic development. It is also contended that the basic model underlying Adam Smith's 'Wealth of Nations' specially refers to farm sector.[6] The early classical economists made the productivity of agriculture the key to the size of country's population and hence to considerable extent to the size of its entire economy. Modern economists also recognized the importance of agriculture. Theodore Schultz wrote,

> There is no longer any room for doubt whether agriculture can be a powerful engine for growth. For acquiring such an engine it is necessary to invest in agriculture and this is not simple because so much depends on the form the investment takes. Incentives to guide and reward farmers are a critical component. Once there are investment opportunities and efficient incentives the farmers will turn sand into gold.[7]

T.W. Schultz has, in his book, *Economic Crisis in World Agriculture*, emphasized the need to increase agricultural productivity,

Eminent Swedish economist, Gunnar Myrdal, had warned of widespread starvation together with the threat for nuclear war as the two most 'menacing calamities' which face mankind in the next several years. The security and peace of the world are threatened by pervasive hunger in developing countries. There is a necessity for much greater emphasis on increasing production on land both to feed growing populations and to feed growing industries. Growth of agriculture has to take place in a right relationship to the other sectors of the economy. Acute shortage of farm commodities will push up their prices and subsequently the prices of other commodities will also rise. W. Arthur Lewis[8] rightly depicted the importance of agriculture as follows : Rising agricultural productivity supports and sustains industrial development in several important ways. Firstly it permits agriculture to release a part of its labour force for industrial development while meeting the food needs of the non-farm sector. Secondly it raises agricultural incomes, thereby creating the rural purchasing power needed to buy the new industrial goods and rural savings which may then be mobilised by direct and indirect means to finance industrial development. Finally, it enables agriculture to supply the major wage good (food) of industrial workers. In economically advanced countries like the United Kingdom, the United States of America and Sweden, "the additional labour required for the relatively more rapid expansion of economic activity in the favourable occupations or industries has been obtained mainly from agricultural sources".[9] In the context of Indian Economy it may not be applicable because of existence of a "large number of unemployed in the urban areas and the relatively fast rate of population growth noticed in the last two decades".[10]

The depressed level of agricultural output will negate the development of other sectors of the economy. Increased agricultural productivity will counterbalance this negative effect. Side by side with the economic development in the country, the demand for agricultural raw materials and foodgrains will also show a steadily increasing trend. Expansion of cotton, jute and oil industries depends mainly on the availability of raw materials which also are supplied by the farm sector.

Benjamin Higgins also recognised the importance of increased agricultural productivity, but his central thesis is "only a rapid change to extensive. mechanised agriculture, with enough industrialisation to absorb the population displaced from the rural sector, will assure take-off into steady growth".[11]

Professor Bauer has very aptly recommended an appreciable advancement of agriculture so as to help the industrialisation of the country. He advocated that it is highly probable, both on general grounds and on Indian experience of the last half a century, that there can be no substantial improvement in the general living standards of people without an appreciable advantage in the huge agricultural sector which has been a drag on the economic development of the country. A substantial increase in agricultural production is necessary to provide an assured surplus for a large industrial population, a market for the output of industry and an appreciable contribution to Government revenues for the financing of essential services. In our country Japanese model of economic growth may be more suitable and the Indian planners have been fully alive to this by according the highest priority to agriculture in all the Five Year Plans except the Second Five Year Plan.

In an underdeveloped country like India the demand for froodgrains increases automatically because of higher income elasticity of demand for foodgrains. Dr. P.V. John advocates,

> If agriculture stagnates, the capitalist profits remain a small part of the national income and savings and investment are accordingly small. Smooth economic development requires that industry and agriculture should grow together.[12]

If a low income country can not solicit foreign aid or foreign commercial investment and thus chooses not to receive foreign aid, it shall have to bank upon its own resources which may chiefly come from the farm sector. Contribution from agriculture to capital formation may come through agricultural taxation. Decline in agricultural prices resulting in increased agricultural production may shift savings and investment to non-farm sectors, perhaps through financial intermediaries like the postal savings in Japan.

In England the spectacular development of the national economy for a century after 1850 was preceded by a very long period of major structural shifts in agriculture, which has been described as the agricultural revolution in economic history. From the British experience one can obviously infer that an agricultural revolution is regarded as a pre-condition for industrialisation and economic development. For today's heavily overpopulated but underdeveloped world this western experience may be of questionable value. However, it does have some universal application and utility. Industrial development is usually unthinkable in the absence of agricultural development. Industrialisation through foreign investments will only lead to economic dualism and will remain patchy and insecure till agriculture is modernized. There can not be an alround economic development in a country where the majority of the people are dependent upon a backward traditional agriculture. The process of modernization of agriculture calls for a change in the techniques of cultivation, organization and character of agriculture.

India is predominantly an agricultural country where roughly 70 per cent of its population has remained engaged in agriculture during the post-independence period. On an average, 50 per cent of national income has been contributed by the agricultural sector during the above-mentioned period. Inspite of such a huge percentage of population engaged in agriculture most of the people are ill-fed and in ill-health by the standards of what man has been able to accomplish in the prosperous nations of the world. In our country there is a need to improve both the quantity and quality of diet. To feed more people a better quality diet, agricultural production has to be increased through increased productivity. The diet taken by an average Indian lacks in required contents. As a result health is impaired, resistance to disease is lowered and the capacity to work is reduced. If human energy is kept so low by malnutrition as to significantly retard the development process an increase in agricultural production and resultant improvements in diet might in itself have a significant effect in increasing production. Improved diet affects human attitudes and aspirations than a purely physical work capacity. Keeping in view the above facts it is clear that agricultural development presents a basis for economic development in India too. The importance of agricultural development in the early stages of economic growth of India obtains from the fact that agriculture is the source of food, and the increase in demand for food in India is related to a rapidly rising population. Demand for food grows with the rise in incomes because of existence of high income elasticity of demand for food in the underdeveloped country like India.

Self-sufficiency in foodgrains is to be achieved by increasing both production and productivity in the agricultural sector. Rising imports of food is not an honourable solution for solving the food problem. Sometimes self-sufficiency in foodgrains becomes vital for the independence and existence of a country. The Government of India could take up a bold and independent stand during the Indo-Pakistan war of 1971 against the direct and indirect pressures of the most powerful nation of the world and win it mainly due to self-sufficiency in foodgrains, brought about by 'Green Revolution', which was the result of use of new high yielding varieties of wheat. Had our country been weak on the food front, it could not ignore the dictates of the donor country. Food exports may be used by a donor country as a leverage to influence our foreign policy or pressurise us into adopting a particular attitude on an international issue, which otherwise food aid receiving country may not like to adopt. The latest thinking in the world politics is that during the next quarter of a century, by the end of which the world population is likely to double itself, food is going to be the greater weapon in the arsenal of the powerful nations of the world.

It is clear that a nation can ignore the subject of food at its own peril, no matter what the level of progress it might have achieved in the other spheres. The highest priority may be given to investment in agriculture because therein lies the highest marginal productivity of capital and the supply of food need be increased so as to cope up with the problem of additional demand for food.

In our country Agricultural Production Team remarked in the year 1959 that an immediate and drastic increase in food production is the primary problem. Without enough food India's hopes for improving human welfare, achieving social justice and securing democracy are simply doomed. It is undoubtedly true that production on land is to be increased to feed growing population and industries. The policies to be adopted to achieve this objective differ in nature. In India, the agriculture will continue to play a decisive role in the achievement of plan targets. The success of planning will definitely depend on the performance of agriculture. The success or failure of agricultural development programmes ultimately depends decisively on the way farmers react to such agricultural development programmes. It is the decision of the farmer which determines the allocation of land resource for various crops.

The importance of agriculture in the context of economic development of our country has been discussed in great detail. Now, a brief discussion on the causes of backwardness of Indian agriculture and the suggestions for increasing the agricultural yield are also presented here. In our country backwardness of agriculture is largely due to institutional deficiencies. Reorganisation of the methods of production and distribution would lead, through better incentives, to greater effort and thus larger incomes. Lack of capital is also responsible for the low level of production in agriculture and it results either from an overall scarcity of capital or the private and public sectors' inability to adequately appreciate the importance of agriculture. Most of the capital formation has to take place in the agricultural sector. Over-population with its menace of disguised unemployment in agriculture is probably one of the most important factors responsible for low productivity and low level of capital formation in the farm sector. Check on population growth is a solution to overcome the difficulties caused by superfluous numbers. Agricultural backwardness arises from lack of diversification of employment. Promotion of industrialisation is a step towards the solution of the problem. There has been a slow response of farm population to new ideas and changes in the methods of resource utilisation. Proper emphasis upon education, innovation, scientific methods of cultivation and the provision of appropriate leadership may also solve the problem. Changes in sociological rigidities can also help attracting farmers to use the modern farm technology.

AREA SELECTED FOR INVESTIGATION

The region selected for investigation is the state of Uttar Pradesh. In terms of area it is the fourth largest state of India, while in terms of population it ranks first. It covers an area of 294,413 square kilometres (9.6 per cent of total area of the country). According to Census 1971 the total population of the State was 88.3 million (16.2 per cent of total national population). The density of population (number of persons per square kilometre) worked out to 300. The density of population in the country and the neighbouring states of Rajasthan, Madhya Pradesh and Haryana were 177, 75, 94 and 227* respectively. It is quite clear that the State is having a heavy pressure of population per square kilometre compared to All India and its neighbouring States. The State is having a preponderant influence on the economic as well as political affairs of the country due to its size, strategic position and being the most populous territory in the country. It is the only State in the country whose boundary line was not affected by the reorganisation of the States that took place in 1956. With such a large segment of population living within its boundaries, the economic development of Uttar Pradesh assumes special significance. By the sheer weight of its population, economic growth in this region would naturally affect the average level of living of the country as a whole.

Uttar Pradesh is a land locked State lying between 23° 52′ and 31° 18′ N and 77° 10′ and 39° 39′ E. It is bounded by Tibet and Nepal in the North, Himachal Pradesh in the north west, Madhya Pradesh in the south and Bihar in east. The Yamuna river forms part of the eastern boundary. Other boundaries are artificial.

Physical features

The State is divided into four regions : the hill areas in the north along the Himalyas ; the sub-montane tract comprising Bhabar and Tarai, the Yamuna-Gangetic Plain. The Himalayan tract in the extreme north comprises the Kumaon Division and part of Dehra Dun district. This region is mountainous with some very high peaks (of which Nanda Devi is the highest 7,698 metres). It nowhere assumes the comparative low level of a plateau and includes some of the wildest and most magnificent country in the whole range of the Himalayas. These mountains have in-

*Census of India, 1971.

fluenced and moulded the State's economy greatly and will continue to affect it in several ways in future. The snow-clad peaks

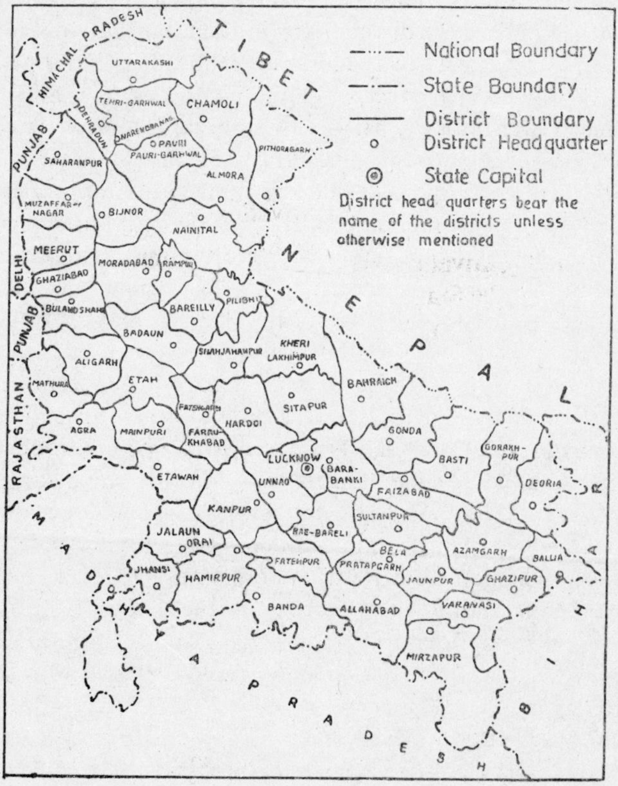

MAP 1 : *Uttar Pradesh : Administrative Division*

give rise to several rivers which carry silt to the plains rendering the land fertile for agriculture. These rivers provide a perennial source of irrigation, sites for generating hydel power, serve as means of transport and promote the development of fisheries. The mountainous region has also several minerals and a large part of the area under forests which could give rise to several industries. The entire State has tropical monsoon climate except in the Himalayan region where the climate is temperate.

IMPORTANCE OF AGRICULTURE IN THE STATE ECONOMY

The economy of the State is predominantly agricultural and that it has not been keeping pace with the economic development of the country. The share of the State in the total national income which was 13.5 per cent in 1960-61 was reduced to 12.9 per cent in 1970-71. In 1970-71 the per capita income of the State was 20 per cent lower than the per capita national income.

The preponderance of agriculture in the State's Economy is discernible from the fact that the contribution of agriculture and allied sectors comprising cultivation, animal husbandry and fishery in the total income of the State is 60 per cent as against only about 50 per cent in the country as a whole. The importance of agriculture in the State can also be judged from the fact that about 86 per cent population of the State resides in rural areas and 75 per cent of the total workers engaged in any gainful activity are employed in agriculture alone either as cultivators or as agricultural labourers. The State is a major producer of sugarcane. The State's agriculture has also contributed significantly by providing raw materials to several agro-based industries such as food processing, grain milling, oil crushing and sugar industries. The dominant position of agriculture in the national context is due more to the size of the State than to the performance of agricultural industry. From the productivity analysis of farm crops it is abundantly clear that the agricultural productivity is low. The State is in an advantageous position regarding agricultural development. The major part of the cultivated area in the State is covered by fertile thick alluvial soil. Normally the State gets a reasonable rainfall. In 1974-75 the area irrigated was 40 per cent of the total net area sown. The role of agriculture in the development of the State economy is vital because it supports a major portion of the State's population and also because there is great scope for further development. In the development of national economy also the role of State's agricultural development has a decisive role to play. Keeping in view magnitude of importance of the State in the development of the country, Uttar Pradesh has been selected for the purpose of present investigation.

EDUCATIONAL ACHIEVEMENT OF THE STATE

A special mention is to be made of the educational achievement in the State. The literacy rate in the rural areas is an index of the cabability of farmers in the State to adopt new farm technology. Table 1·1 summarises the achievement of the State in the attainment of literacy as revealed by Census of 1971.

TABLE 1·1 : **Literacy in Uttar Pradesh—1971**

Category	Literate as per cent of total population		
	Rural	Urban	Total
1. Male	28.02	52.08	31.50
2. Female	6.99	33.33	10.55
3. Persons	18 13	43.63	21.70

The literacy rate in the State is 21.70 per cent which is much less than the All India literacy rate of 29.45 per cent. The literacy rate in the State is much higher in urban areas compared to rural areas. The lowest level of literacy is found among rural females. Among the rural females the level of literacy is 6.99 per cent only. The rate of literacy among rural males is 28.02 per cent whereas the literacy rate among urban males is 52.08 per cent. There is an urgent need to raise the rural literacy rate. It will indirectly raise the receptiveness of farmers to improved farm technology.

OBJECTIVES OF THE INVESTIGATION

The main objectives of the investigation are :

(*i*) study of the broad trends in area, output and yield of the main crops viz, rice, wheat, sugarcane, barley, gram and maize in the State during 1950-51 to 1975-76;

(*ii*) analysis of behaviour of farm (harvest) prices and whole sale prices of the crops mentioned above ;

(*iii*) study of supply response of rice, wheat and sugarcane ; and (*iv*) projections of demand of foodgrains in the State on the basis of present availability.

It is obligatory to point out that the concept of productivity as used here represents production raised per unit of land. In other words, productivity represents yield of crops per hectare. Productivity of other inputs used in the production of crops falls beyond the scope of the study. The study covers the analytical and graphical survey of trends in area, output and yield of rice, wheat, sugarcane, barley, gram and maize. Impact of changes in prices on acreage allocation among crops forms an important part of the study. During the period of investigation farm technology underwent a change, availability of credit facilities in the rural areas increased due to the fast expansion of banking facilities in the rural areas, mass media like radio and television brought to the door step of the farmers the latest farm technology and the rate of literacy increased considerably in the rural areas due to fast expansion of educational facilities. The study of the supply response of the concerned crops in the State during the period when the above mentioned changes took place will be thought-provoking for planners and administrators alike. Demand projections worked out on the basis of present availability would give an insight into the magnitude of the effort to be made for increasing the foodgrains production keeping in view the fast rate of growth of population in the State.

PLAN OF THE STUDY

The investigation has been spread over seven chapters. Each chapter deals with a specific issue. A brief discussion on the role of agriculture in the economic development of the country has been already made in this introductory chapter. Besides, a layout plan and objectives of the investigation have been given in this chapter.

In the second chapter trends in area, production, and yield of selected crops have been studied both analytically and graphically. This chapter also includes a discussion on the quality and the sources of data used in the study.

The chapter third studies the changes in the cropping pattern in the State during the period 1950-51 to 1974-75. Growth rates of area and production of the selected crops have been also worked out for intensive study.

The fourth chapter deals with structure and behaviour of the prices of selected farm crops. It is well known that the farmers while making any change in investment in agriculture have to look for its profitability which ultimately depends on price or productivity increase. This chapter includes a discussion on the trends in farm (harvest) prices and wholesale prices of selected agricultural commodities.

In the chapter fifth the variables used in the Nerlovian Model have been discussed. Also, it throws light on the Supply Response Coefficients obtained by applying Nerlovian Supply Response model to the data pertaining to selected crops. The policy implications of the coefficients so obtained have also been discussed. A comparative study of these results has been made with the results obtained in the earlier studies.

Chapter sixth deals with the demand projections of foodgrains through 1961. These demand projections have been worked out on the basis of per capita availability of foodgrains and population projections of the State prepared by the Expert Committee, Registrar General of India, the Ministry of Home Affairs, New Delhi.

The concluding chapter, besides dealing with the summary of the study, includes a discussion on some of the policy implications emerging from the study.

In the end appendix tables and a selected bibliography has been given.

REFERENCES

1. First Five Year Plan, Planning Commission, Government of India, New Delhi, p. 44.
2. Second Five Year Plan, Planning Commission, Government of India, New Delhi, p. 259.
3. Third Five Year Plan, Planning Commission, Government of India, New Delhi.
4. United Nations : Process and Problems of Industrialisation in underdeveloped countries, p. 3.
5. Coale and Hoover, Population Growth and Economic Development in Low Income Countries : A Case Study of India's Prospects, pp. 120-139.

6. Sir John Hicks, *Capital and Growth*, 1965, ch. iv.
7. T.W. Schultz, *Transforming Traditional Agriculture*, Yale University Press, *1965*.
8. W. Arthur Lewis, *Theory of Economic Growth*, London George Allen and Unwin, 1955.
9. E.M. Ojala, *Agriculture and Economic Progress*, Oxford University Press, 1952, p.85.
10. A.M. Khusro, Economic Development with No Population Transfers (a working paper prepared at the Institute of Economic Growth, Delhi), Asia Publishing House, Bombay, 1962.
11. Benjamin Higgins, *Economic Development*, p. 455.
12. P V. John, *Some Aspects of the Structure of Indian Agricultural Economy, 1947-48 to 1961-62*, Asia Publishing House, New Delhi.

2

Study of Trends of Area, Production and Yield of Selected Crops

IN THIS chapter an attempt is made to study the broad trends and variations in area, production and yield of selected crops in the State. The time period chosen for analysis of these trends is post-independence period 1950-51 to 1975-76. Needless to say that the study is bound to be highly aggregative as it relates to a State which consists of such subdivisions which differ in their geographical characteristics, aptitudes of the people and cropping pattern. Moreover, the study is based on a time series data for a period of twenty five years. During the course of this time period the procedural techniques of statistical data collection have undergone changes varying from crude eye estimation methods including hunches to highly sophisticated techniques of statistical data collection.

An analysis of the broad trends of yield per hectare (productivity) on the basis of historical time series is useful in planning of perspective agricultural development in the State. Statistical analysis of the behavioural pattern of area, production and yield of the selected crops gives an insight into effectiveness of various economic forces operating in the State. Now the State has been released partially from the subsistence character and considerable changes have occurred in the demand for farm commodities.

The chapter concentrates mainly on the study of trends in area, production and yield of rice, wheat, sugarcane, barley, gram and maize. Maize and sugarcane; gram, barley and sugarcane;

rice and wheat; are respectively taken as the competitive crops of rice, wheat and sugarcane. For correct estimation of the trends in area, production and yield it is necessary to have a continuous series of data on the area, yield and production crops on a comparable basis. The official estimates of area, yield and production of principal crops published from time to time are not comparable mainly due to improvements made in their quality during the last two and a half decades. An observation made by the Directorate of Economics and Statistics substantiates our argument. According to the above mentioned organ of the Ministry of Agriculture and Irrigation, "the improvements relate mainly to the substitution of the traditional eye estimation methods by the objective random sampling crop cutting methods and a gradual expansion of the reporting area. In the process, while the quality of the estimated figures improved from year to year a strict comparability over time has been lost".[1]

Data

It is not possible to collect reasonably accurate data directly from the farmers for so many years past. Therefore, the research worker has to depend entirely on the secondary data collected by various agencies of governments, semi-governments and research organisations. The quality of data used in the analysis has a direct bearing on the derived conclusions. Therefore it will not be out of place to briefly discuss the nature and quality of the data used. The data collected during post-independence period are supposed to be more reliable compared to the data relating to pre-independence period. The realism of this assumption is based on the fact that several improvements and refinements have been made in the collection of statistical data. Scientific and objective statistical techniques were deployed for collection of data with the development of modern statistical machinery. The statistical machinery has been equipped with highly sophisticated statistical techniques of data collection pertaining to measurement of economic and social characteristics. Obsolete and subjective methods such as Eye Estimation Method and Hunches have been replaced by more scientific and objective methods of statistical data collection. In the pre-independence period the collection of statistical data emerged as the by-product of administrative machinery. During the post-independence period the statistical data have been collected with the specific objective of economic planning in view and it has resulted in improvements in the quality of data.

The data requirement of the present investigation falls into four categories : Crop areas, crop yields, crop production and crop prices. Discussion on the crop prices is reserved for the fourth chapter. Areas sown to crops are more faithful reflectors of farmers' decisions than areas matured or harvested. The trends in area throw light on the variability of the area of crops under cultivation. The variance in area allocation reveals the extent to which the farmers are lured by economic forces in allocating their land to different crops. The physical volume of output traces the behavioural trends of output during the period 1950-51 to 1975-76. The trends in crop areas and physical volume of output are realistic indicators of development of the crops covered in the investigation. An analysis of historical time of yield gives an insight into the growth pattern of productivity of crops considered in the study.

The principal source of data relating to crop areas, crop yields and crop production is official blue book *Estimates of Area, Production and Yield of Principal Crops in India,* Directorate of Economics and Statistics, Ministry of Agriculture and Irrigation, Government of India, New Delhi. The methods of data collection adopted by the above mentioned organ of the Central Government have been subjected to severe criticisms in the past.[1]

With the passage of time these criticisms have been gradually met. A mention may be made of the different methods of data collection which were in use at the beginning of the period selected for the investigation. The following methods were adopted by the agencies responsible for data collection :

(*i*) the annawari or percentage estimation ;

(*ii*) the direct estimation ; and

(*iii*) estimation through random sample crop cutting surveys.

These methods varied according to the procedures followed for estimating the average yield per hectare.

In all the above mentioned methods, the framing of a crop estimate involved determining the acreage under the crop, its average yield per hectare and finally its volume of output. Out of these variables, the area under the crops was fairly accurate.[2] This was so because cadastral data were available in all the temporary revenue settlement areas

As regards the second variable involved in estimation of volume of output, namely, the average yield per hectare, different procedures were in vogue with regard to the different methods mentioned above. In the annawari method, the average yield was calculated by the following formula :

Average yield per hectare=Standard (normal) yield[3] per hectare multiplied by the season factor. Mere estimates provided the basis for seasonal or conditional factor. Keeping in mind the ill-defined standard yield, the patwari (lekhpal) estimated the proportion of the current crop to the standard yield. The standard yield was expressed in terms of sixteen annas in the State and the current crop was estimated in terms of annas it was better or worse than the standard. This seasonal or conditional factor was averaged all over the villages of the district. The district officer could modify the so obtained average on the basis of his personal knowledge or opinion. The subjectivity involved in estimation of such crop yields calls for cautious interpretation of the results.

In the direct estimation method, the average yield per hectare was obtained directly from the standard yield without bringing in the condition factor. The accuracy of these estimates were not very different from the annawari estimates because these estimates were also based on eye estimates of village officials which also involved subjectivity.

The method of random crop cutting surveys was introduced with the purpose of removing arbitrariness and errors inherent in the methods mentioned above. This method was not adopted simultaneously[4] in the State in respect of selected crops. In the year 1949-50 this method was adopted for the estimation of production for the crops of wheat, barley and gram. In the year 1950-51 the crop cutting method was adopted for the estimation of production of rice and maize. In the year 1960-61 the estimation of production of sugarcane came under the method of crop cutting surveys. A random sample crop cutting method is an objective procedure for estimating average yield. This is a tested standard procedure free from personal bias and can be relied upon to provide estimates with a fair degree of precision. This procedure has been followed in the State of Uttar Pradesh only gradually. Average yields per hectare of principal crops have been obtained by dividing the total production by the corresponding area under each crop. The average yields per hectare have generally been calculated on the basis of area and production figures rounded up to hundreds. Production of sugarcane is given in terms of *gur*. The statistical figures of area, production and yield relate to the agricultural year which runs from July to June.

AGRICULTURAL DEVELOPMENT

Earlier a mention of this fact has been made that the State consists of several agro-climatic subdivisions which differ in soil fertility, topography, altitudes, water supply and climatic conditions. These differences result in great variations in farming patterns followed and productivity levels achieved. However, instead of examining the performance of the subdivisions separately a review of overall changes in area production and yield of selected crops has been considered. The study of these growth patterns with reference to agriculture assumes special importance in a planned economy because it helps in locating the weaknesses in the existing programmes of agricultural development.

Growth pattern of area, production and yield of selected crops

In the main, three types of data are treated *viz*., the volume of output, area and yield. Three yearly moving averages of volume of output and yield have been calculated for all the selected crops. This device was employed to eliminate the random variations in the data and to get a clear picture of the year to year growth rates of output in the State. In Table 2.1, year to year area variations of selected crops are presented.

Rice

Three crops of rice namely, summer rice, autumn rice and winter rice are raised in the State. The area under rice includes the

TABLE 2.1 : Annual area variation of selected agricultural crops in Uttar Pradesh : 1951-52 to 1977-78

(*'000 hectares*)

Year/Crop	Rice	Wheat	Sugarcane	Barley	Gram	Maize
(1)	(2)	(3)	(4)	(5)	(6)	(7)
1. 1951-52	−360.9	+ 43.8	+186.3	− 61.0	−118.1	− 9.9
2. 1952-53	+ 89.4	+ 16.1	−129.5	− 56.5	+360.3	+ 36.3
3. 1953-54	+ 83.1	+359.4	−271.9	+114.5	− 13.8	+161.5
4. 1954-55	+ 15.1	+ 26.7	+128 9	−111.8	− 10.0	+ 27.8
5. 1955-56	+ 83.7	+270 9	+173.2	+ 23.7	+ 59.5	− 89.9
6. 1956-57	+ 94.3	+ 6.1	+140.2	+ 27.4	−299.7	+ 54 5
7. 1957-58	+ 39.6	−291.4	− 14.1	−210.7	+ 86 0	+ 69 0
8. 1958-59	+233.8	+123.2	−115.1	+ 82.1	+148 5	+ 10.1
9. 1959-60	+ 18.0	+ 5.9	+ 79.7	− 12.3	− 28.1	− 1.6
10. 1960-61	+ 36.2	+ 62.4	+137.6	+ 0.5	−141.4	− 37.6
11. 1961-62	− 10.5	+161.2	+ 33.7	− 22.3	+ 23.5	+ 31.9
12. 1962-63	+104.0	− 61.0	−100.6	−147.0	−108.0	− 11.0
13. 1963-64	+ 71.0	−122.0	− 33.0	− 94 3	+132.0	+ 11.4
14. 1964-65	+100.3	+ 48.4	+160.0	−105.7	+ 27.1	+ 34.1

15.	1965-66	−242.5	+149.4	+100.7	+ 4,8	− 43.4	+ 49.5
16.	1966-67	+239.0	+279.5	+300.0	+ 28.3	−200.2	+ 91.8
17.	1967-68	− 47.5	+575.4	−192.7	+102.2	−120.2	+194.2
18.	1968-69	+122.8	+269.4	+205.9	−201.1	− 18 4	+ 46.1
19,	1969-70	+ 12·8	+139.3	+173.7	+ 73.7	+ 16.6	+ 3.2
20.	1970-71	+ 28.6	+528.8	− 31.4	−161.8	−183.6	+ 2 9
21.	1971-72	+159.7	+138.7	− 71.1	− 11.0	− 89 3	− 29.2
22.	1972-73	−347.8	+ 89.4	+ 33.9	− 24.1	− 68.1	+ 3.7
23.	1973-74	+ 96.5	−126.2	+164.7	− 19.9	+ 35.4	+ 0.5
24.	1974-75	+ 58.9	+142.9	+ 18.8	− 36.1	−234.8	− 88.9
25.	1975-76	+ 92.3	+150.1	− 50.8	− 29.7	+ 4.4	+331.2
26.	1976-77	+ 30.5	+321.9	+ 15,2	−127.1	− 66.8	− 72.6
27.	1977-78*	+175.4	+351.9	+235.5	−100.0	− 26.6	− 71.8

* Provisional.

Source : Computed from Appendix Table 2.1.

areas under all the crops of rice irrespective of the season of the year in which it is grown. During the period 1950-51 to 1976-77, the area under rice decreased in five years *viz.*, 1951-52, 1961-62, 1965-66, 1967-68 and 1972-73. The highest decrease in area under the crop was noticed in 1951-52 (360.9 thousand hectares) followed by 1972-73 (347.8 thousand hectares) and the lowest decrease in area under the crop took place in 1961-62 (10.5 thousand hectares) preceded by 1967-68 (47.5 thousand hectares). The highest increase in area was noticed in 1966-67 (239.0 thousand hectares) followed by 1958-59 (233.8 thousand hectares) and the lowest increase in area under the crop took place in 1969-70 (12.8 thousand hectares) preceded by 1954-55 (15.1 thousand hectares). The area under the crop increased by 20.8 per cent during 1950-51 to 1976-77.

Wheat

The area under wheat decreased in 4 years out of 26 years. The highest decrease in area under the crop was noticed in 1957-58 (291.4 thousand hectares) followed by 1973-74 (126.2 thousand hectares) and the lowest decrease in area was observed in 1962-63 (61.0 thousand hectares) preceded by 1963-64 (122.0 thousand hectares). The highest increase in area under the crop took place in 1967-68 (575.4 thousand hectares) followed by 1970-71 (528.8 thousand hectares). The lowest increase in area under the crop took place in 1959 60 (5.9 thousand hectares) preceded by 1956-57 (6.1 thousand hectares). The area under the crop increased by 99.8 per cent during 1950-51 to 1976-77.

Sugarcane

The area under sugarcane decreased in 10 years out of a time period of 26 years. The highest decrease in area of the crop was noticed in 1953-54 (271.9 thousand hectares) followed by 1967-68 (192.7 thousand hectares). The lowest decrease in area under the crop took place in 1957-58 (14.1 thousand hectares) preceded by 1970-71 (31.4 thousand hectares). The highest increase in area under the crop took place in 1966-67 (300.0 thousand hectares) followed by 1968-69 (205.9 thousand hectares). The lowest increase in area under the crop took place in 1976-77 (15.2 thousand hectares) preceded by 1974-75 (18.8 thousand hectares).

The overall position is that the area under the crop increased by 43.6 per cent during 1950-51 to 1976-77.

Barley

The area under barley decreased continuously in the time periods 1961-62 to 1964-65 and 1970-71 to 1976-77. The area under the crop decreased in 16 years out of a total of 26 years. The highest decrease in area under the crop was noticed in 1957-58 (220.7 thousand hectares) followed by 1968-69 (201.1 thousand hectares) and the lowest decrease in area under the crop took place in 1971-72 (11.0 thousand hectares) preceded by 1973-74 (19.9 thousand hectares). The highest increase in area under the crop took place in 1953-54 (114.5 thousand hectares) followed by 1967-68 (102.1 thousand hectares) and the lowest increase in area took place in 1960-61 (0.5 thousand hectares) preceded by 1965-66 (4.8 thousand hectares). The overall position is that the area under the crop decreased by 44.8 per cent during 1950-51 to 1976-77.

Gram

The area under gram decreased in 16 years out of a time period of 26 years. The highest decrease in area under the crop was noticed in 1974-75 (234.8 thousand hectares) followed by 1956-57 (229.7 thousand hectares). The lowest decrease in area under the crop took place in 1973-74 (0.5 thousand hectares) preceded by 1954-55 (10.0 thousand hectares). The highest increase in area under the crop was in 1952-53 (360.3 thousand hectares) followed by 1958-59 (148.5 thousand hectares) and the lowest increase in area under the crop took place in 1975-76 (4.4 thousand hectares) preceded by 1969-70 (16.6 thousand hectares). The area under the crop decreased by 32.0 per cent during 1950-51 to 1976.77.

Maize

The area under maize showed a decrease in 8 years out of a time period of 26 years. The highest decrease in area under the crop was noticed in 1955-56 (82.9 thousand hectares) followed by 1974-75 (88.9 thousand hectares) and the lowest decrease in area under the crop took place in 1959-60 (1.6 thousand hectares)

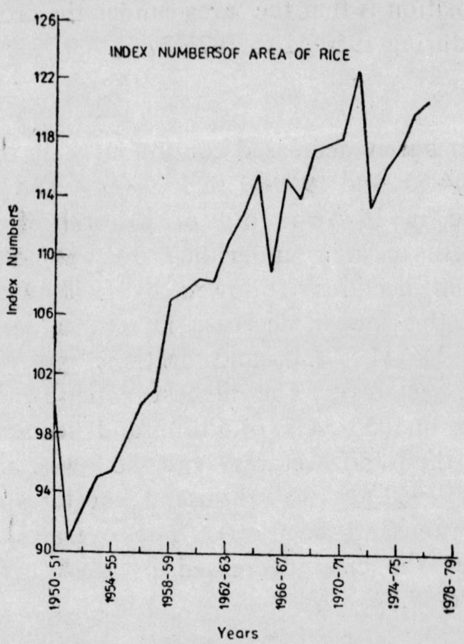

GRAPH 2A

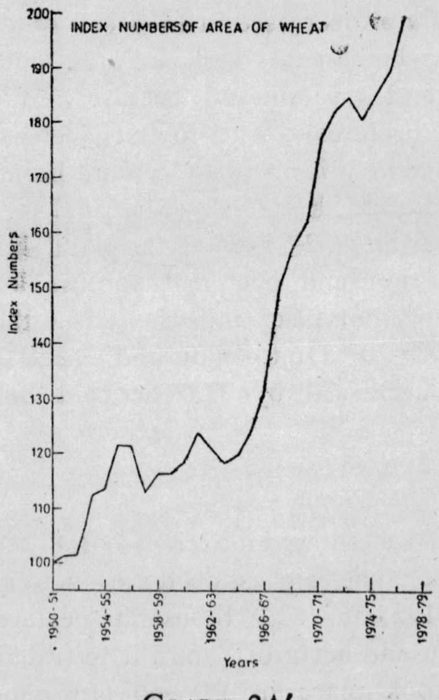

GRAPH 2A′

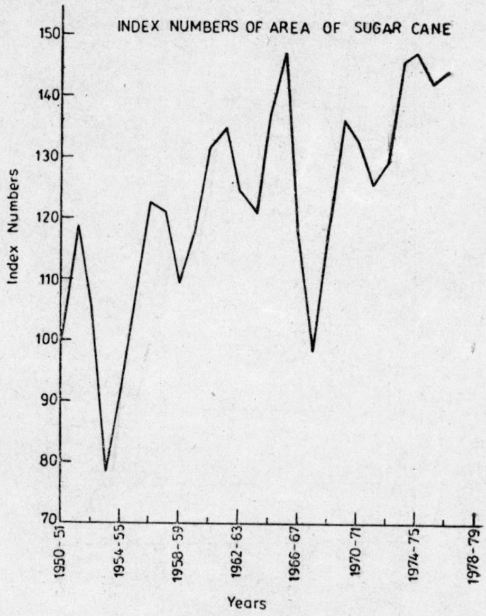

GRAPH 2A″

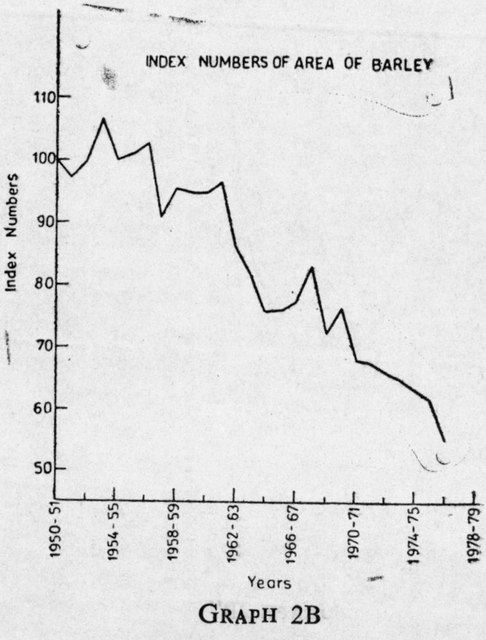

GRAPH 2B

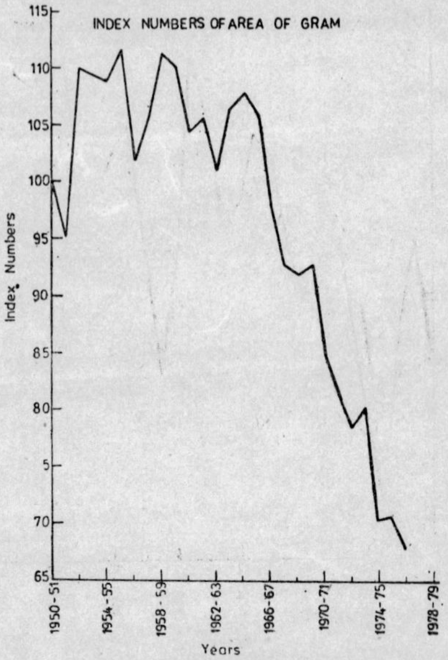

GRAPH 2B′

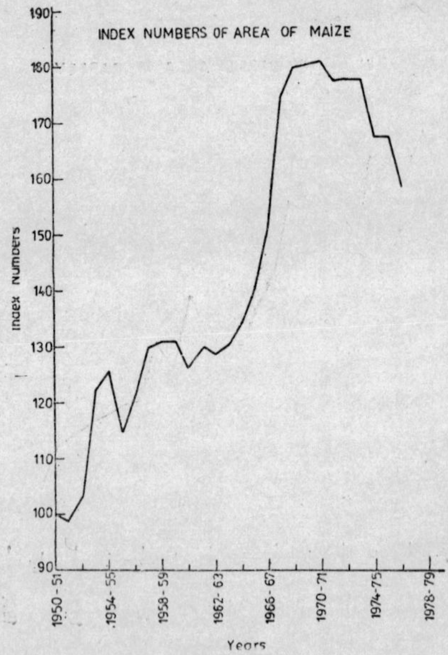

GRAPH 2B″

TABLE 2.2 : **Index numbers of area of selected agricultural crops in Uttar Pradesh : 1950-51 to 1976-77**

Year/Crop	Rice	Wheat	Sugarcane	Barley	Gram	Maize
(1)	(2)	(3)	(4)	(5)	(6)	(7)
1. 1950-51	100.0	100.0	100 0	100.0	100.0	100 0
2. 1951-52	90.6	101.3	118.3	96.9	95.2	98.8
3. 1952-53	93.0	101·8	105.5	99.8	109.9	103.2
4. 1953-54	95.1	112.6	78.7	105.7	109.4	122.5
5. 1954-55	95.5	113.5	91.4	99.9	109.0	125 9
6. 1955-56	97.7	121.6	108.6	101.1	111.4	115.1
7. 1956-57	100.1	121.8	122.4	102.5	102.0	121.6
8. 1957-58	101.2	113.0	121.0	91.2	105.5	129.9
9. 1958-59	107.2	116.7	109.7	95.4	111.6	131 1
10. 1959-60	107.7	116.9	117.5	94.8	110.4	130.9
11. 1960-61	108.6	118.8	131.1	94.8	104 6	126.4
12. 1961-62	108.4	123.7	134.4	96.7	105·6	130.2
13. 1962-63	111.1	121.8	124.5	86.1	101.2	128.9
14. 1963-64	112.9	118.1	121.2	81.3	106.6	130.3
15. 1964-65	115.5	119.6	137.0	75.8	107.7	134.4

(1)	(2)	(3)	(4)	(5)	(6)	(7)
16. 1965-66	109.2	124.1	146.9	76.1	105 9	140.3
17. 1966 67	115.4	132.5	117.4	77.5	97.7	151.3
18. 1967-68	114.2	149.9	98.3	82.8	92.8	174.6
19. 1968-69	117.4	158.0	118.6	72.5	92.0	180.1
20. 1969-70	117.7	162.0	135.8	76.3	92.7	180.5
21. 1970-71	118.4	178.2	132.6	67.9	85.2	180.9
22. 1971-72	122.6	182.4	125.6	67.4	81.5	177.4
23. 1972-73	113.6	185.0	129.0	66.1	78.7	177.8
24. 1973 74	116.1	181.2	145.3	65.1	80.2	177.9
25. 1974-75	117.6	185,5	147.1	63.3	70.6	167.2
26. 1975-76	120.0	190.1	142.1	61.7	70.8	167.4
27. 1976-77	120.8	199.8	143.7	55.2	68.0	158.6

Source : Computed from Appendix Table 2.1

preceded by 1951-52 (9.9 thousand hectares). The highest increase in area under the crop took place in 1976-77 (331.2 thousand hectares) followed by 1967-68 (194.2 thousand hectares) and the lowest increase in area took place in 1973-74 (0.5 thousand hectares) preceded by 1970-71 (2.9 thousand hectares). The overall position is that the area under the crop increased by 58.6 per cent during 1950-51 to 1976-77.

Index numbers of area under selected crops

The index numbers of area under the selected crops *viz.*, rice, wheat, sugarcane, barley, gram and maize are presented in Table 2.2. The analytical inferences of Table 2.1 are fully corroborated by the index numbers given in Table 2.2.

The Table 2.3 presents the percentage change in area in 1976-77 over 1950-51.

TABLE 2.3 **Percentage change in area in 1976-77 over 1950-51**

Crop	Rice	Wheat	Sugarcane	Barley	Gram	Maize
(1)	(2)	(3)	(4)	(5)	(6)	(7)
Percentage Change	+20.8	+99.8	43.7	−44.8	−32.0	+58.6

Source : Table 2.2.

The graphs 2A's and 2B's respectively depict the movements of area under rice, wheat, sugarcane ; and barley, gram and maize.

Growth of production

After studying the growth pattern of area under the selected crops, the growth of output of these crops is taken up for discussion.

Quantitative indications of changes in output of the selected crops can be had from Table 2.4 and Table 2.5. In columns 2 and 3 of table 2.4 are given the average outputs of crops during the trienniums 1950-51 to 1952-53 and 1973-74 to 1975-76. Attributing the three yearly average output to the middle of each period the estimate of average change in output of each crop is obtained. It is seen from the Table 2.4 that the output of rice crop in 1975-76 (the mid year of the ending trien-

TABLE 2.4 : **Annual average trend rates of production judged by two-three year averages, 1950-51 to 1952-53 to 1974-75 to 1976-77 of selected crops**

Crop	Average production		Change in production between 1950-53 to 1974-77		Annual Average Change	
	1950-51 to 1952-53	1974-75 to 1976-77	Quantity	Percentage	Quantity	Percentage
(1)	(2)	(3)	(4)	(5)	(6)	(7)
1. Rice	1810	4036	+2226	+123.0	+92.8	+5.1
2. Wheat	2721	8223	+5502	+202.2	+229.3	+8.4
3. Gur	2954	6169	+3215	+108.8	+134.0	+4.5
4. Barley	1672	1303	− 369	−22.1	−15.4	−0.9
5. Gram	1448	1207	− 241	−16.6	−10.0	−0.7
6. Maize	721	979	+ 258	+35.8	+10.8	+1.5

Source : Computed from Appendix Table 2.2.

nium) showed an increase of 123 per cent over that in 1951-52 (the mid year of the ending triennium). It means an average annual increase of 5.1 per cent. The output of wheat, sugarcane (gur) and maize showed percentage increase of 229.3, 134.0 and 10.8 respectively. It means an average annual percentage increase of 8.4, 4.5 and 1.5 respectively for the crops of wheat, sugarcane and maize. During the period mentioned above the output of barley and gram showed a decrease of 15.4 per cent and 10.0 per cent which means an average annual decrease of 0.9 per cent and 0.7 per cent respectively. The broad conclusion of the above analysis is that the output of two crops viz., barley and gram showed a decrease and the output of other four crops showed an increase during the entire period of investigation. The Table 2.5 presents year to year growth rates of output of the selected crops. A brief survey of the trend growth of output of these crops is given in table 2.6.

Rice

The highest (11.2%) and the lowest (0·4%) positive rates of growth of output occurred in the years 1953-54 and 1955-56 respectively. The highest (14.2%) and the lowest (0.2%) negative rates of growth of output were observed in the years 1965-66 and 1963-64 respectively. The overall rate of growth of output in 1976-77 compared to 1950-51 worked out to be 114.7 per cent.

Wheat

The highest (16.9%) and the lowest (0.1%) positive rates of growth of output occurred in the years 1967-68 and 1955-56 respectively. The highest (10.9%) and the lowest (0.1%) negative rates of growth of output took place in the years 1962-63 and 1957-58 respectively. The overall growth of output in 1976-77 compared to 1950-51 worked out to be 228.5 per cent.

Sugarcane (Gur)

The highest (24.5%) and the lowest (.01%) positive rates of growth of output occurred in 1959-60 and 1974-75 respectively. The highest (12.0%) and the lowest (0.7%) negative rates of growth of output were noticed in the years 1966-67 and 1970-71

TABLE 2.5 : **Annual rates of growth of output of selected agricultural commodities in Uttar Pradesh, 1952-53 to 1976-77**

Year/crop	Rice	Wheat	Gur	Barley	Gram	Maize
(1)	(2)	(3)	(4)	(5)	(6)	(7)
1. 1952-52	+ 5.41	+ 5.33	− 9.51	+1.44	+ 7.25	+ 0.97
2. 1953-54	+11.16	+ 8.90	− 4.30	+4.60	+10.95	+ 5.49
3 1954-55	+11.13	+ 2.34	+ 3.83	−3.16	− 2.55	− 8 33
4. 1955-56	+ 0.42	+ 0.09	+18.15	−3 14	− 4.71	+15 48
5. 1956-57	+ 2.07	− 6.10	+ 2.61	+9.25	− 7.56	− 0.62
6. 1957-58	+ 6.13	− 0.07	+ 1.43	−4. 64	+ 7.23	− 0.25
7. 1958-59	+ 1.83	+ 1.43	− 2.82	−3.13	− 0.44	+ 1.24
8. 1959-60	+10.38	+13.08	+24.48	+9.68	+ 8.23	− 6.99
9. 1960 61	+ 3.64	+10.37	+16.88	+7.39	− 5.27	+ 3.16
10. 1961-62	+ 7.47	− 0.74	+ 7.54	−3.10	+ 0.80	− 1.[illegible]0
11. 1962-63	+ 1.31	−10.88	− 4.67	−1.21	− 8.33	+ 6 61
12. 1963 64	− 0.22	− 0.36	+ 3.51	−7.72	+ 9.16	+ 7.90
13. 1964-65	− 8.14	+ 5.41	+ 9.41	+0.23	+ 0.37	+ 4.62
14. 1965-66	−14.16	+14.31	− 5.10	+4.40	+11.04	+10.66
15. 1966-67	− 0.78	+14.25	−12.04	+8.22	− 5.05	+ 7.98

16.	1967-68	+ 7.60	+16.86	− 4.55	−4.38	− 0.41	+ 6.58
17.	1968-69	+18.56	+12.90	+16.67	+4.50	− 3.72	+ 3.30
18.	1969-70	+ 4.51	+10.13	+11.25	−7.67	− 2.11	+18.00
19.	1970-71	+ 8.42	+ 7.29	− 0.72	+5.19	+ 0.43	−11.44
20.	1971-72	− 2.34	+ 5.57	− 2.40	−3.95	+16.45	+ 4.31
21.	1972-73	+ 1.45	− 7.96	− 3.81	−6.46	−10.96	−17.72
22.	1973-74	− 2.31	− 1.78	+ 7.26	+0.16	−13.63	− 0.46
23.	1974-75	+ 9.57	+ 5.05	+ 0.01	+1.96	− 6.06	−10.54
24.	1975-76	+ 3.70	+14.16	+ 0.02	neg.	+ 9.62	+ 0.03
25.	1976-77	+13.38	+ 9.46	+10.41	−6.91	+ 5.22	+ 3.88

Source : Computed from Appendix Table 2.2

TABLE 2.6 : **Three yearly moving averages of volume of output of selected crops**

(000' tonnes)

Year/Crop	Rise	Wheat	Sugarcane (Gur)	Barley	Gram	Maize
1. 1951-52	1810	2721	2954	1672	1448	721
2. 1952-53	1908	2866	2673	1696	1553	728
3. 1953 54	2121	3121	2558	1774	1723	768
4. 1954-55	2357	3194	2656	1718	1679	704
5. 1955-56	2367	3197	3138	1664	1600	813
6. 1956-57	2416	3002	3220	1510	1479	808
7. 1957-58	2564	3000	3266	1440	1586	806
8. 1958-59	2611	3043	3174	1395	1579	816
9. 1959 60	2882	3441	3951	1530	1·09	759
10. 1960-61	2987	3798	4618	1643	1620	783
11. 1961-62	3210	3770	4966	1592	1632	772
12. 1962-63	3252	3360	4734	1399	1496	823
13. 1963-64	3245	3348	4900	1291	1633	888
14. 1964-65	2981	3529	5348	1294	1639	929
15. 1965-66	2559	4034	5075	1351	1820	1028
16. 1966-67	2539	4609	4464	1462	1728	1111
17. 1967-68	2732	5386	4261	2398	1721	1183

18.	1968-69	3239	6081	4971	1461	1657	1222
19.	1969-70	3385	6697	5530	1349	1622	1442
20.	1970-71	3670	7185	5490	1419	1629	1277
21.	1971-72	3584	7585	5358	1363	1524	1332
22.	1972-73	3636	6981	5562	1275	1357	1096
23.	1973-74	3552	6857	5966	1277	1172	1091
24.	1974-75	3892	7203	6020	1302	1101	976
25.	1975-76	4036	8223	6169	1303	1207	979

Source : Computed from Appendix Table 2.3

respectively. The overall growth of output in 1976-77 compared to 1950-51 worked out to be 121.1 per cent.

Maize

The highest (18.0%) and the lowest (0.003%) positive rates of growth of output occurred in 1969-70 and 1975-76 respectively. The highest (−17.7%) and the lowest (−0.3%) negative rates of growth took place in 1974-75 and 1957-58 respectively. The overall increase of output in 1976-77 compared to 1950-51 worked out to be 68.5 per cent.

Barley

The highest (9.7%) and the lowest (0.2%) positive rates of growth of output were achieved in 1959-60 and 1973-74 respectively. The highest (6.5%) and the lowest (1.2%) negative rates of growth took place in the years 1974-75 and 1962-63 respectively. The overall decrease in output in 1976-77 compared to 1950-51 worked out to be 31.8 per cent.

Gram

The highest (11.0%) and the lowest (0.4%) positive rates of growth of output occurred in 1965-56 and 1964-65 respectively. The highest (13.6%) and the lowest (0.4%) negative rates of growth were observed in 1973-74 and 1967-68 respectively. The overall decrease in output of gram in 1976-77 compared to 1950-51 worked out to be 6.3 per cent.

Growth of yield

Most of our land is already under cultivation. The scope for horizontal expansion of cultivable land in the State is quite limited except for marginal increases through the improvement of degraded soil or the provisions of drainage in water logged areas. The only way out to increase the production in the State is through the vertical expansion of land. The vertical expansion of land is possible through the adoption of improved farm technology. Here the main thrust is to increase yield per hectare. The machinery of the government responsible for development of agriculture in the State has made every effort to create requisite facilities for the application of improved farm technology. A review of the growth pattern of yields of important selected crops in the State is enough for making an assessment of success achieved by the government in increasing the yields of crops.

In Table 2.8 and Table 2.9 are presented respectively the three yearly moving averages and index numbers of yields of important selected crops *viz.*, rice, wheat, sugarcone, barley, gram and maize. Moving averages have been calculated to obtain trend values so that fluctuations in the time series of yields are eliminated or in any case reduced to minimum. Trend values of yields so obtained reveal the general tendency of the time series of yields. A study of broad trends of yields of the selected crops is attempted below :

Rice

In the base year 1950-51 the average yield of rice per hectare was 519 kilogrammes and it increased to 922 kilogrammes per hectare in the year 1976-77. The average yield per hectare in the first triennium 1950-51 to 1952-53 was 496 kilogrammes and it increased to 875 kilogrammes per hectare in the last triennium ending 1976-77. The average rate of growth of yield between the two trienniums worked out to be 76.6 per cent. The index numbers of yield revealed an increase of 77.7 per cent in 1976-77 compared to 1950-51. Column 2 of table 2.8 presents the three yearly moving averages of yields of rice. The smoothed series and index numbers of yields indicate an increasing behaviour of yield of rice during the entire period of investigation.

A clear view of the movements of yields of rice is depicted in Graph 2C.

Wheat

In the base year 1950-51 the average yield of wheat per hectare was 821 kilogrammes and in the year 1976-77 it increased to 1350 kilogrammes per hectare, thereby showing an increase of 64.4 per cent during the entire period of 1950-51 to 1976-77. The average yield in the first triennium 1950-51 to 1952-53 and the last triennium ending in 1976-77, worked out to be 812 and 1291 kilogrammes respectively. The rate of growth of average yields between the two trienniums worked out to be 59 per cent.

TABLE 2.7 : **Index numbers of production of selected agricultural crops in Uttar Pradesh, 1950-51 to 1976-77**

Year/Crop	Rice	Wheat	Sugarcane (Gur)	Barley	Gram	Maize
(1)	(2)	(3)	(4)	(5)	(6)	(7)
1. 1950-51	100.0	100.0	100.0	100.0	100.0	100.0
2. 1951-52	77.5	94.6	109.2	88.1	89.0	106.4
3. 1952-53	94.2	105.4	91.3	104.7	109.9	125.6
4. 1953-54	114.3	116.0	71.4	104.3	121.7	103.4
5. 1954-55	109.6	122.6	97.4	101.7	124.9	125.0
6. 1955-56	129.5	113.6	101.3	95.0	100.8	96.1
7. 1956-57	116.2	116.3	120.5	94.7	105.3	153.6
8. 1957-58	117.0	101.1	105.7	74.7	99.1	122.4
9. 1958-59	151.7	113.4	106.0	82.8	122.8	95.4
10. 1959-60	123.3	121.1	111.0	86.7	103.9	158.1
11. 1960-51	157.6	145.0	184.8	98.5	136.0	96.0
12. 1961-62	167.3	152.7	173.8	102.5	104.4	106.6
13. 1962-63	156.7	118.0	146.5	77.8	106.4	153.3
14. 1963-64	164.0	99.8	161.2	64.7	97.9	119.5
15. 1964-65	166.3	151.3	190.7	83,6	132,6	136.4

16.	1965-66	117.2	138.0	192.0	78.4	107.7	172.2
17.	1966-67	100.7	155.5	133.5	74.6	135.3	165.3
18.	1967-68	163.2	214.7	128.5	103.1	113.6	174.2
19.	1968-69	146.2	223.7	171.3	67.1	106.3	205.5
20.	1969-70	176.7	232.0	205.7	85.7	122.2	183.1
21.	1970-71	185.2	282.3	185.3	83.5	106.2	275.9
22.	1971-72	189.0	277.5	167.3	79.4	107.8	129.4
23.	1972-73	163.8	276.2	192.3	75.9	100.5	208.4
24.	1973-74	193.1	216.1	206.0	68.1	71.8	167.1
25.	1974-75	176.3	263.7	208.4	79.8	69.5	127.1
26.	1975-76	214.8	314.3	197.8	80.2	86.0	155.7
27.	1976-77	214.7	328.5	221.1	68.2	93.7	168.5

Source : Computed from Appendix Table 2.2.

TABLE 2.8 : **Three yearly moving averages of yields of selected agricultural crops in Uttar Paradesh, 1951-52 to 1976-77**

Year/Crop	Rice	Wheat	Sugarcane (Gur)	Barley	Gram	Maize
(1)	(2)	(3)	(4)	(5)	(6)	(7)
1. 1951-52	496	812	2703	868	583	858
2. 1952-53	532	820	2611	864	605	817
3. 1953-54	582	861	2750	895	646	795
4. 1954-55	636	833	2816	863	627	695
4. 1955-56	628	812	2892	844	611	805
6. 1956-57	630	761	2709	786	572	792
7. 1957-58	645	772	2741	765	611	764
8. 1958-59	643	794	2703	763	592	749
9. 1959-60	694	883	3221	827	645	701
10. 1960-61	716	955	3538	894	622	725
11. 1961-62	763	936	3763	890	644	720
12. 1962-63	763	834	3686	819	588	761
13. 1963-64	745	842	3780	822	636	812
14. 1964-65	686	881	3907	860	629	822

No.	Year						
15.	1965-66	586	971	3720	910	582	867
16.	1966-67	584	1017	3638	952	580	606
17.	1967-68	614	1100	3771	921	608	840
18.	1968-69	722	1170	4137	966	734	821
19.	1969-70	745	1213	4225	961	739	958
20.	1970-71	797	1242	4115	1035	772	851
21.	1971-72	786	1259	4092	1042	764	892
22.	1972-73	804	1150	4113	988	694	740
23.	1973-74	796	1123	4195	1013	627	747
24.	1974-75	857	1167	4100	1057	615	685
25.	1975-76	876	1291	4217	1113	711	716
26.	1976-77	972	1379	4435	1118	760	769

Source : Computed from Appendix Table 2.3.

TABLE 2.9 : **Index numbers of yield of selected agricultural crops in Uttar Pradesh, 1950-51 to 1976-77**

Year/Crop	Rice	Wheat	Sugarcane (Gur)	Barley	Gram	Maize
(1)	(2)	(3)	(4)	(5)	(6)	(7)
1. 1950-51	100.0	100.0	100.0	100.0	100 0	100 0
2. 1951-52	85.6	93.3	92.2	90.9	93.5	107.7
3. 1952-53	101.4	103.5	86.5	104.9	100.0	121.8
4. 1953-54	120.4	102.9	90.6	98.6	111.2	83.4
5. 1954-55	114.6	108.0	106.5	101.7	113.9	99.2
6. 1955-56	132.6	93.3	93.3	93.9	90.4	83.5
7. 1956-57	116.0	95.5	98.4	92.3	103.4	126.5
8. 1957-58	115.6	89.4	87.6	81.9	94.4	94.4
9. 1958-59	141.4	97.1	96.7	86.7	110.1	72.7
10. 1959-60	114.5	103.5	94.5	91.5	94.1	120.7
11. 1960-61	145.1	121.9	141.1	103.9	120.3	75.9
12. 1961-62	154.5	123.4	129.4	109.3	98.8	81.8
13. 1962-63	141.3	96.8	117.6	90.3	105.2	118.8
14. 1963-64	145.3	84.4	133.0	79.6	92.0	91.7

15.	1964-65	143.9	126 4	139.1	110.2	123.0	101.5
16.	1965-66	107.3	111.1	130.7	103.3	101.7	122.7
17.	1966-67	87.3	117,3	113.7	96.8	68.1	108.8
18.	1967-68	143.0	143.1	130.1	124.6	122.3	99.7
19.	1968-69	124.3	141.5	144.4	92.6	115.4	114 1
20	1969 70	150.1	143.0	151.5	112.3	131.7	101.4
21.	1970-71	156.3	158.6	139.7	122.8	124.7	152.5
22	1971 72	154 2	152.1	133.1	117.7	132.2	73.0
23.	1972-73	144.1	149.2	149.0	114.7	127.7	117.2
24.	1973-74	166.3	119.1	141.8	104.6	89.6	94.0
25.	1974-75	149.9	142.0	141.6	126.0	98.5	75 9
26.	1975-76	179.0	165.3	139.2	129.9	121.5	93.1
27.	1976-77	177.7	164.4	153.9	123.4	137.8	106.2

Source : Computed from Appendix Table 2.3.

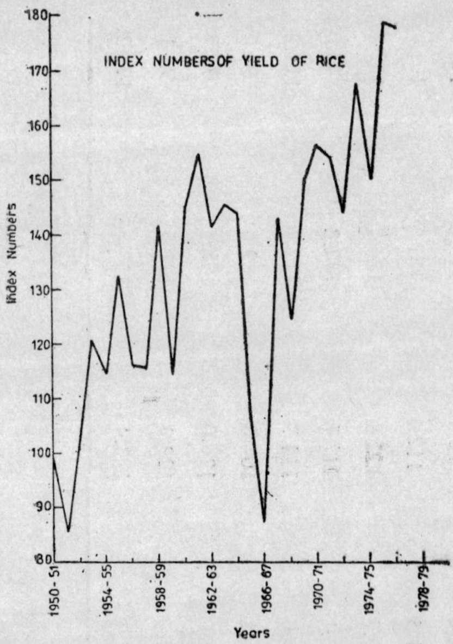

GRAPH 2C

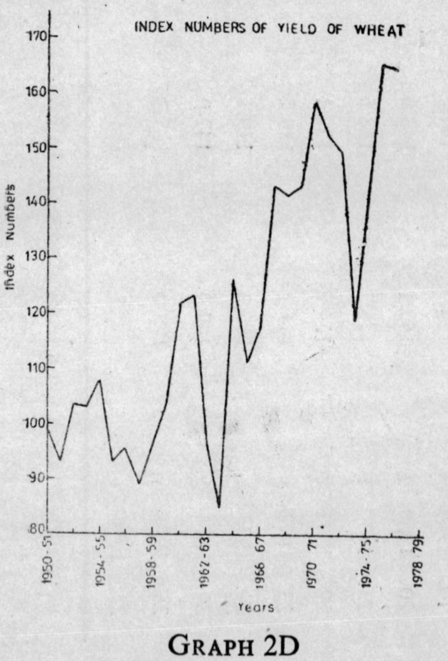

GRAPH 2D

The index number of yield in 1976-77 increased to 164.4 compared to 1950-51=100. Therefore, the rate of growth in yield in 1976-77 over 1950-51 worked out to be 64.4 per cent. Column 3 of tables 2.8 and 2.9 present three yearly moving averages and index numbers of yields of wheat respectively. The smoothed series and index numbers of yield of wheat indicate an increasing trend.

A clear view of the movements of yields of wheat is depicted in Graph 2D.

Sugarcane (*Gur*)

In the base year 1950-51 the average yield of sugarcane in terms of gur was 2910 kilogrammes per hectare and it increased to

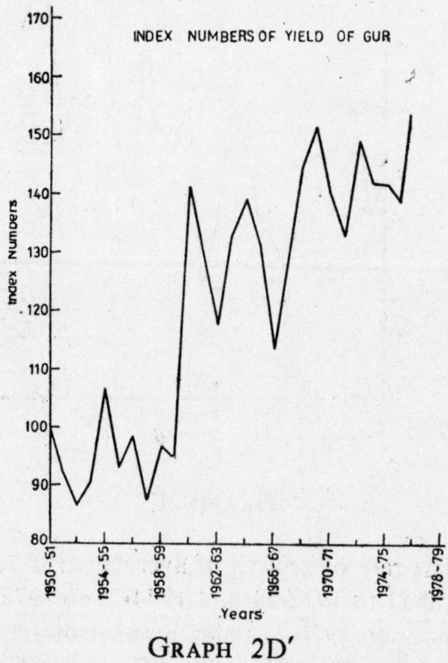

GRAPH 2D′

4479 kilogrammes per hectare in 1976-77. The rate of increase in average yields between the two trienniums worked out to be ˂6 per cent. In the year 1976-77 the index number of yield of gur increased to 153.9 showing thereby, an increase of 53.9 per cent over the base year 1950-51. Column 4 of tables 2.8 and 2 9 present three yearly moving averages and index numbers of yield of gur respectively. The index numbers of yield indicate an increasing trend.

A clear view of the movements of yield of gur is depicted in Graph 2D′.

Barley

The yield of barley per hectare in the base year 1950-51 was 880 kilogrammes and it increased to 1086 kilogrammes per hectare

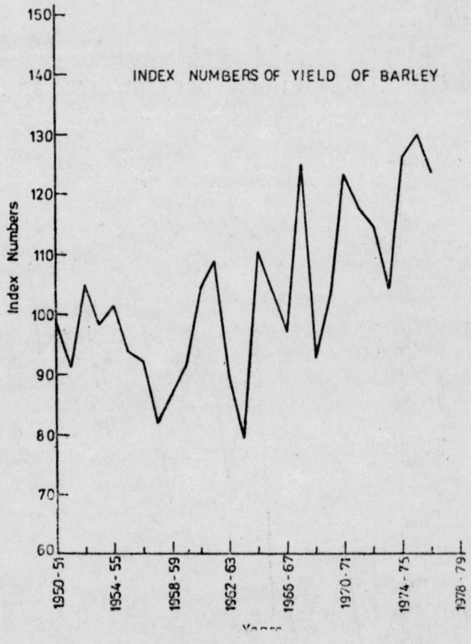

GRAPH 2E

in 1976-77. The rate of growth of average yield between the two trienniums 1950-51 to 1952-53 and 1974-75 to 1976-77 worked out to be 28 per cent. In 1976-77 index number of yield of barley was 123·4 compared to 1950-51=100. This rate of increase in yield of barley is much lower than the rate of increase in yields of superior foodgrains like rice and wheat. An emerging conclusion of this phenomena is that the programmes of agricultural development in the State have concentrated more on increasing the yields of superior foodgrains. The yield of barley ranged between 700 and 1143 kilogrammes per hectare. The three yearly moving average and index number series of yields of the crop show a mixed trend.

A clear view of the behavioural pattern of yield is depicted in Graph 2E.

Gram

In the base year 1950-51 the average yield of gram was 596 kilogrammes and it increased to 821 kilogrammes in 1976-77. The rate of increase in average yields between the first and the last trienniums worked out to be 22 per cent. The rate of increase in yield in 1976-77 over 1950-51 worked out to be 37.8 per cent. The yield of gram ranged between 406 and 821 kilogrammes per hectare during the entire period of investigation.

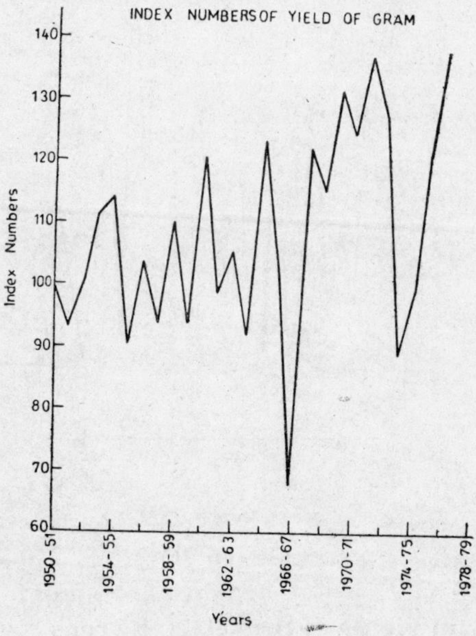

GRAPH 2F

The Graph 2F depicts the movements of the yield of gram during the period of investigation.

Maize

The average yield of maize per hectare was 781 kilogrammes and it rose to 829 kilogrammes in the year 1976-77. The rate of increase in average yield between the first and the last trienniums worked out to be 17 per cent. The index number of yield in 1976-77 was 106.2 showing thereby an increase of 6.2 per cent in yield of maize in the year 1976-77 compared to the base year 1950-51. The yield per hectare ranged between 568 and 1191 kilogrammes respectively.

The Graph 2G clearly depicts the movements of yields of maize during the entire period of investigation.

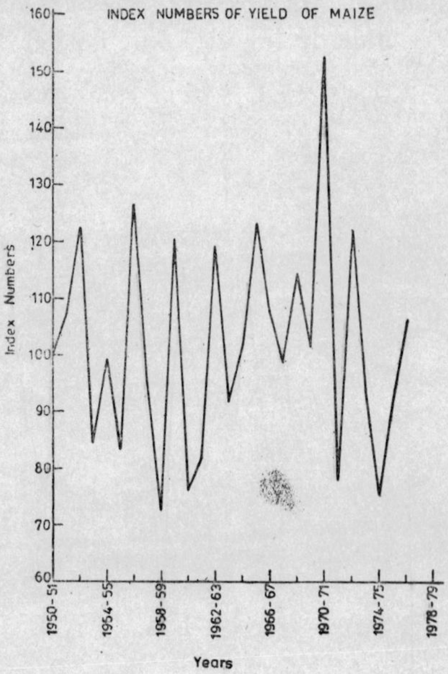

GRAPH 2G

Variability in yields of the selected crops

Comparative Variability of yields of the selected crops is presented in Table 2.10.

From column 2 of table 2.10 it is noticed that gur had the highest average yield (3530 kilogrammes) per hectare and the lowest average yield (645 kilogrammes) per hectare was obtained in case of gram. From column 4 it is seen that the highest

TABLE 2.10 : **Variability of yields of selected crops, 1950-51 to 1976-77**

S.No.	Crop	Mean	Standard Deviation	Coefficient of Variation
	1	2	3	4
1.	Rice	694.2	128.7	18.5
2.	Wheat	987.2	201.2	20.4
3.	Gur	3530.2	651.1	18.4
4.	Barley	817.7	117.6	14.4
5.	Gram	644.8	96.2	14.9
6.	Maize	792.3	146.6	18.5

Source : Computed from Appendix table 2.3.

variability (20.4 %) was present in the yield of wheat and the lowest variability (14.4 %) was observed in yield of barley. The variability in yields of rice, gur, maize and gram was 18.5, 18.4, 18.5 and 14.9 per cent respectively. The variability present in the yields of these crops ranged between 14 and 20 per cent. It may be added here that the coefficient of variation is the appropriate statistical measure to compare the variability present in the yields of the concerned crops.

GROWTH RATES OF AREA, PRODUCTION AND YIELD OF SELECTED CROPS

The index numbers of area, production and yield of the selected crops *viz.*, rice, wheat, sugarcane, barley, gram and maize have been utilised to calculate the growth rates by employing the following exponential function :

$$I=AB^t \qquad \text{...}(i)$$

I Index Number under the category

t time

A constant (equal to at I the base level when $t=0$) and

$B=1+\frac{r}{100}$ where r is the percentage rate of growth of area/production/yield of the particular crop.

Taking logarithms, function (1) reduces to the following linear function with log I as dependent variable and t as independent variable

$$\text{Log } I=a+bt$$

where

$$a = \log A \text{ and } b = \log B$$

Percentage rate of growth per annum

$$r = (B-1)\ 100$$
$$= (\text{Antilog } b - 1)\ 100$$

which represents a uniform rate of change from observation to observation.

A summary statement of the growth rates obtained on the basis of applying linear and exponential models to the time series data of area, production [and yield of selected crops is presented below :

TABLE 2.11 : **Growth rates of area, production and yield of various groups in U.P., 1950-51 to 1976-77**

Crops	Linear $Y=a+bt$		Exponential $Y=ab^t$	
	Coefficient	r^2	Coefficient	r^2
AREA				
(*i*) Rice	1.097	.864	.0046	.853
(*ii*) Wheat	3.764	.868	.0114	.891
(*iii*) Sugarcane	1.573	.505	.0058	.476
(*iv*) Barley	(−) 1 821	.914	−.0098	.908
(*v*) Gram	(—) 1.417	.673	−.0068	.671
(*vi*) Maize	3.081	.819	.0096	.834
PRODUCTION				
(*i*) Rice	4.043	.745	.0123	.7098
(*ii*) What	8.512	.808	.0206	.8394
(*iii*) Sugarcane (Gur)	4.902	.757	.0150	.7437
(*iv*) Barley	(−) 1.002	.404	−.0051	.3958
(*v*) Gram	(−) 0.644	.101	−.0030	.1204
(*vi*) Maize	3.241	.372	.0097	.4199
YIELD				
(*i*) Rice	2.329	.534	.0076	.4923
(*ii*) Wheat	2.602	.684	.0092	.6675
(*iii*) Sugarcane (Gur)	2.495	.753	.0093	.7454
(*iv*) Barley	1.170	.438	.0047	.4100
(*v*) Gram	.876	.179	.0032	.1393
(*vi*) Maize	.052	.0005	.00008	.C0006

Source : **Computed from the Appendix tables 2.1, 2.2, 2.3.**

On examination of growth rates presented in table 2.11 it is quite evident that the area, production and yield of four out of six selected crops increased during the entire period of investigation. While the rates of growth of area and production of barley and gram were negative, the growth rates of yield of both these crops were positive. The decline in area and production of barley was more pronounced compared to decline in area and production of gram. The highest positive rate of growth in area and production was seen in case of wheat but in yield it lagged behind the rate of growth of yield of sugarcane. In case of yield of maize the growth rate was very low.

One emerging conclusion is that the area under superior foodgrains is on the increase while the area under inferior cereals like barley is on the decrease.

REFERENCES

1. Estimates of Area, Production and yield of principal Crops in India. Directorate of Economics and Statistics, Ministry of Agriculture and Irrigation, New Delhi, 1971.
2. Report of the Royal Commission on Agriculture in India, London, 1928, p. 605; Bowley and D.H. Robertson, A Scheme for an Economic Census in India, New Delhi, 1934, p. 35 ; P.J. Thomas and N. Sundaram Sastry, Indian Agricultural Statistics, Madras, 1939, p. 35; K.T. Shah and K.J. Khambata, Wealth and Taxable Capacity of India, Bombay, 1924, pp. 85-6 : Guide to Current Agricultural Statistics, Ministry of Food and Agriculture, Delhi, 1954, pp. 12-15; Proceedings of Board of Agriculture, p. 31 (1919) and pp. 49-51 (1924).
3. Recent Advances in Collection, Analysis, and Interpretation of Agro-Economic Data in India, *Studies in Agricultural Economics* ; Vol. II, Ministry of Food and Agriculture, Government of India, New Delhi, 1956, p. 146.
4. The standard yield per hectare is the yield of a standard crop divided by the number of hectares on which it is grown. The standard crop is officially defined as "that crop which past experience has shown to be the most generally recurring crop in a series of years, the typical crop of the local area, the crop which the cultivator has a right (as it were) to expect and with which he is (or should be) content and if less he has reason to complain". Cited from "Manual on Preparation of Crop Forecasts", India, Development of Statistics, 1917., p. 5.
5. Estimates of Area, Production and yield of Principal Crops in India, Directorate of Economics and Statistics, Ministry of Agriculture and Irrigation, New Delhi, 1971.

3

Changes in Cropping Pattern and Land Utilisation

THIS CHAPTER is divided into two sections. Section I deals with the study of changes in cropping pattern. Section II deals with (*i*) the changes in land utilization ; (*ii*) growth of fertilizer distribution and area under high yielding varieties ; (*iii*) distribution of rainfall and growth of assured irrigation.

SECTION I

Review of changes of cropping pattern gauges the shifts in areas under different crops over a period of time. The heterogeneity and possibility of crop substitution are the two important characteristics of agricultural land which deserve special mention in studies pertaining to cropping pattern changes. Heterogeneity arises from agro-climatic differences existing within the various regions of the State. The heterogeneity of land also arises due to differences in availability of irrigation facilities among different regions of the State. Investigations pertaining to study of changes in cropping pattern assume special importance in taking cognizance of soil-climate factors and the crops that could be grown within a particular environment.[1] Impact of changes in technological, economic and institutional factors can be felt only when the existing cropping pattern undergoes a change. Generally, the farmers have a tendency to stick to a stable cropping pattern under any given agro-climatic region and they do not shift much from this position except to the extent dictated by price factors in adjusting acreage allocations. Breakthrough in farm technology makes a room for new production opportunities which do change the cropping pattern but with some time lag. Newer production opportunities become possible by a new technology such as use of high yielding crop varieties along with chemical fertilizers, pesticides, insecticides and increase in irrigated area. In the State of Uttar Pradesh, as in other States of the country, the severe constraints in increasing the land area and pressure of growing population reflected in growing aggregate demand, has made it inevitable that further increase in agricultural output can come through land productivity and intensive land utilization.[2] Mention of this fact has been made in chapter 2 by pointing out that the possibilities of horizontal expansion of land are limited and the only way out is to increase the productivity of land by vertical expansion.

As the study is based on aggregate State level data, it does not take into account the heterogeneity factor and the disaggregated crop substitution possibilities within regions or districts of the State.

The cropping pattern of individual or group of farmers within each district determine the aggregate cropping pattern. In allocation of land resource the farmers maximise their expected return. The State cropping pattern reflects the rational decisions of the aggregate of farmers subject to technical and institutional constraints.

The constraints imposed by the nature of subsistence farming practised by the majority of farmers in the State needs special mention.

The farmers respond to changes in input costs, output prices and crop yields in their acreage decisions but they are reluctant to make changes in the cropping pattern on a large scale. Lack of sufficient credit facilities, ignorance and uncertain future prospects restrain the farmers from undertaking any significant changes in their area allocations. The slower rate of adoption of high yielding varieties programme by small farmers in the State is a sufficient proof of this pattern of behaviour. In the long run the changes in cropping pattern do occur. In the light of this hypothesis the study of change in cropping pattern of the State is undertaken.

Changes in cropping pattern in Uttar Pradesh

A review of changes in aggregate cropping pattern in the State during the period 1950-51 to 1974-75 is briefly presented. The quinquennial average area under different crops and the relative share of each crop in gross cropped area has been deployed for the study of cropping pattern. The table 3.1 presents these details of the State with reference to the time period 1950-51 to 1974-75.

Cropping pattern : 1950-51 to 1954-55

The position in 1950-51 to 1954-55 was that out of gross cropped area of 20289.0 thousand hectares, 3653.4 thousand hectares were under rice, 3509.5 thousand hectares were under wheat, altogether contributing 35.3 per cent of gross cropped area. The area under total pulses was 4363.4 thousand hectares or 21.5 per cent of the gross cropped area. The area under total oil seeds was 2289.3 thousand hectares which constitutes 11 per cent of the gross cropped area. The area under jowar was 968.5 thousand hectares (4.8%), barley 1955.9 thousand hectares (9.6%), gram 2553.5 thousand hectares (12.6%). The percentage shares of gross cropped area under rape and mustard, other pulses, sugarcane, maize, small millets, tur and linseed are 6.6, 6.0, 4.9, 4.5, 3.4, 3.2 and 1.6 respectively.

Changes in cropping pattern, 1955-56 to 1959-60

The gross cropped area increased to 21386.0 thousand hectares during the quinquennium 1955-56 to 1959-60. The area under rice increased from 3653.4 thousand hectares to 3959.0 thousand hectares registering slightly upward shift in relative area allocation from 18.0 per cent to 18.5 per cent. An upward increase in the area under wheat from 3509.5 thousand hectares to 3913.1 thousand hectares was observed during the two quinquenniums 1950-51 to 1954-55 and 1955-56 to 1959-60. The area under jowar declined from 968.5 to 790.6 thousand hectares thereby registering a decline in the relative share in gross cropped area from 4.8 per cent to 3.7 per cent. The relative share of barley in gross cropped area decreased to 8.8 per cent from 9.6 per cent. The area under sugarcane increased to 1174.2 thousand hectares,

TABLE 3.1. **Cropping pattern in Uttar Pradesh, 1950-51 to 1974-75**

('000 hectares)

S. No.	Crop	1950-51 to 1954-55	1955-56 to 1959-60	1960-61 to 1964-65	1965-66 to 1969 70	1970-71 to 1974-75
(1)		(2)	(3)	(4)	(5)	(6)
1.	Rice	3653.4	3959.0	4280.0	4421.2	4532.0
		(18.01)	(18.51)	(19.51)	(19.70)	(19.63)
2.	Jowar	968.5	790.6	879.2	841.6	701.2
		(4.77)	(3.70)	(4.01)	(3.75)	(3 05)
3.	Bajra	1114.6	1094.8	1031.9	1026.5	1043.5
		(5.49)	(5.12)	(4.70)	(4.57)	(4.54)
4.	Maize	917.9	1048.4	1084.4	1379 0	1469.5
		(4.52)	(4.90)	(4.94)	(6.15)	(6.39)
5.	Ragi	190.6	192.5	178.9	194.3	239.8
		(0.94)	(0.90)	(0.82)	(0.87)	(1.04)
6.	Small Millets	678.9	627.5	579.7	557.2	476.0
		(3.35)	(2.93)	(2.64)	(2.48)	(2.07)
7.	Wheat	3509.5	3913.1	3992.0	4819.3	6049 9
		(17.30)	(18.30)	(18.20)	(21.48)	(26.31)
8.	Barley	1955.9	1889.2	1681.4	1500.1	1284.7
		(9.64)	(8.83)	(7.67)	(6.69)	(5.59)

Table 3.1 Contd.

(1)	(2)	(3)	(4)	(5)	(6)
9. Total Cereals	14343.6 (70.70)	15001.4 (70.15)	13963.6 (63.66)	14739.2 (65.69)	15796.5 (68.71)
10. Gram	2553.5 (12.59)	2638.8 (12.34)	2564.7 (11.69)	2347.4 (10.46)	1932.8 (8.41)
11. Tur	651.7 (3.21)	635.1) (2.97)	636.5 (2.90)	626.8) (2.79)	536.3 (2.33)
12. Other Pulses	1212.4 (5.98)	1303.0 (6.09)	1411.1 (6.43)	1214.2 (5.41)	1008.1 (4.38)
13. Total Pulses	4363.4 (21.51)	4578.2 (21.41)	4702.3 (21.49)	4188.4 (18.67)	3476.8 (15.12)
14. Total Foodgrains	17353.5 (85.53)	18171.4 (84.97)	18284.6 (83.37)	18927 6 (84.35)	19273.7 (83.83)
15. Groundnut	105.7 (0.52)	168.5 (0·79)	273.8 (1.25)	381.1 (1.70)	350.3 (1.52)
16. Sesamum	510.7 (2.52)	465.0 (2.17)	694.8 (3.17)	674.9 (3.01)	656.3 (2 85)
17. Rape Seed and Mustard	1342.2 (6.62)	1478.6 (6.91)	1856.2 (8.46)	1982.5 (8.84)	2062.2 (8.97)
18. Linseed	329.1 (1.62)	400.5 (1.87)	697.4 (3.12)	732.0 (3.25)	675.9 (2.24)

19. Total Major Oil seeds	2289.3 (11.28)	2520.0 (11.30)	3510.1 (11.44)	3771.6 (16.81)	3745.7 (16.29)
20. Cotton	62.0 (0.31)	68.2 (0.32)	75.6 (0.34)	59.3 (0.26)	44.6 (0.19)
21. Sannhemp	74.6 (0.37)	84.5 (0.39)	67.9 (0.31)	61.9 (0.28)	53.7 (0.23)
22. Potato	91.3 (0.45)	105.4 (0.49)	120.6 (0.54)	155.3 (0.69)	180.8 (0.79)
23. Sugarcane	1002.0 (4.94)	1174.2 (45.9)	1314.2 (5.99)	1251.2 (5.58)	1378.3 (6.00)
24. Gross Cropped Area	20289.0 (100.00)	2138.6 (100.00)	21933 (100.00)	22439 (100.00)	22991 (100.00)

Note : Figures in parentheses are percentages to gross cropped area.

Source : 1. Directorate of Economics and Statistics, Ministry of Agriculture and Irrigation, Government of India.

(*i*) Estimates of Area, Production and Yield of Principal Crops in India.

(*ii*) *Indian Agricultural Statistics*, Vol. I, Summary Tables.

thereby registering an increase in the relative share in the gross cropped area from 4.9 per cent to 5.5 per cent.

Changes in cropping pattern, 1960-61 to 1964-65

During the quinquennium 1960-61 to 1964-65 the gross cropped area increased to 21933.0 thousand hectares. The area under rice increased to 4280.0 thousand hectares and its share in gross cropped area increased to 19.5 per cent. The area under wheat increased from 3913.1 to 3992.0 thousand hectares but, there was a marginal decline in its relative share in gross cropped area from 18.3 per cent to 18.2 per cent. The area under barley decreased from 1889.2 to 1681.4 thousand hectares leading to a decline in its relative share in the gross cropped area from 8.8 per cent to 7.7 per cent. A marginal decrease in area under bajra, ragi, gram and tur was observed in the quinquennium.

Changes in cropping pattern, 1965-66 to 1969-70

During the quinquennium 1965-66 to 1969-70 the gross cropped area increased to 22439.0 thousand hectares. The area under rice increased to 4421.2 thousand hectares, thereby leading to a marginal increase of 0.2 per cent in its relative share in the gross cropped area. Compared to the preceding quinquennium the area under wheat increased by 827.3 thousand hectares, thereby registering an increase of 3.3 per cent in its relative share in the gross cropped area. The area under maize increased by 294.6 thousand hectares and the relative share of the crop in gross cropped area increased from 4.9 per cent to 6.2 per cent. The area under potato crop increased by 34.7 thousand hectares. The area under sugarcane decreased by 63.0 thousand hectares. Thus its percentage share in the gross cropped area decreased from 6.0 per cent to 5.6 per cent. The area under jowar, bajra, small millets, barley, gram, tur and other pulses decreased marginally.

Changes in cropping pattern, 1970-71 to 1974-75

In the last quinquennium the gross cropped area registered an increase of 552.0 thousand hectares over the preceding quinquennium. A major share of this increase in gross cropped area was claimed by wheat and its area increased by 1230.6 thousand hectares, thereby increasing its relative share from 21.5 per cent to 26.3 per cent. The area under sugarcane increased by 127.1 thousand hectares and its relative share in gross cropped area increased marginally from 5.6 per cent to 6.0 per cent. The area under bajra, maize, rape seed and mustard also increased marginally. The area under gram, barley and jowar decreased to the extent of 414.6, 215.4 and 140.4 thousand hectares respectively.

During the entire time period of twenty-five years. 1950-51 to 1974-75, the average gross cropped area increased by 2702.0 thousand hectares, the area under rice increased by 878.6 thousand hectares, that is, its share in increase of gross cropped area was 32.5 per cent, the area under wheat increased by 2540.4 thousand hectares and its share in gross cropped area increase was 94 per cent. There were significant increases in area under maize, tur, groundnut, rape and mustard, linseed, potato and sugarcane. The area under jowar, bajra, small millets, barley, gram, total pulses, cotton and sannhemp decreased considerably.

The emerging conclusion is that the share of rice in gross cropped area ranged between 18 and 20 per cent of gross cropped area whereas the relative share of wheat in gross cropped area increased to 26.3 per cent from 17.3 per cent. The impact of high yielding varieties programme seems to have been more pronounced in wheat than in other crops. The area under inferior foodgrains decreased during the period under review.

The overall picture of cropping pattern that emerges for the State during the period 1950-51 to 1974-75 is that foodgrains occupied over 83 per cent of the gross cropped area with another 16 per cent under oilseeds. Wheat has grown in importance because its percentage share in gross cropped area has gone upto 26.3 per cent whereas the percentage share of rice in the gross cropped area has remained almost static. The relative share of potato crop has gone up to almost one per cent.

SECTION II

Land use pattern in Uttar Pradesh : 1950-51 to 1975-76

Study of statistical data pertaining to different land use categories gives an idea of changes in land use pattern in the State over a period of time. The relevant data on land use pattern for every quinquennium starting with the year 1950-51 and ending with the year 1975-76 is presented in Table 3.2.

In column 1 of Table 3.2 the land use categories given are as follows :

1. Area under land utilisation, 2. Forests, 3. Barren and unculturable waste, 4. Land put to non-agricultural uses, 5. Culturable waste, 6. Permanent pastures and other grazing land, 7. Land under miscellaneous tree crops and groves, 8. Current fallows, 9. Other fallow land, 10. Net area sown, 11. Area sown more than once, 12. Gross cropped area.

At serial No. 13 of column 1 the cropping intensity* is presented. In columns 2 to 8 the data regarding the different land use categories mentioned in column 1 is presented for the years 1950-51, 1955-56, 1960-61, 1965-66, 1970-71 and 1975-76. It is mentioned in the footnote of the Table 3.2 that the data presented do not include the information relating to Kumaon and Uttarakhand Divisions. The data available for the year 1975-76 related to the State as a whole and it was adjusted for the data pertaining to hill portion of the State according to conventional[3] estimates of land use statistics.

During the period 1950-51 to 1975-76 there was an increase of 1.4 per cent in area under land utilisation. The share of area under forests increased from 5.2 per cent to 9.4 per cent. The share of barren and unculturable waste land decreased from 6 per cent to 3 per cent. The definitions of land under current fallows and other fallows had undergone a change in the year 1951-52 and again in 1964-65. Due to these changes in their definitions, the figures for current fallows and other fallows for different years are not strictly comparable but taken together they do give some idea about the changes in the fallow land over years. Both types of fallow lands taken together increased their share in area under land utilisation from 5.3 per cent to 6.1 per cent during the time period 1950-51 to 1975-76. The share of area under non-agricultural uses ranged between 7.0 and 8.5 per cent. The net area sown ranged between 62 and 66 per cent. There was an increase of 53.8 per cent in area sown more than once and the gross cropped area also increased by 16.2 per cent. The cropping intensity increased from 123.4 per cent to 134.0 per cent during

* Cropping intensity is obtained by expressing gross cropped area as per-cent of net cropped area.

TABLE 3.2 : **Land Use* Pattern in Uttar Pradesh, 1950-51 to 1975-76**

('000 hectares)

S.No : Land Use Category	1950-51	1955-56	1960 61	1965-66	1970-71	1975-76
(1)	(2)	(3)	(4)	(5)	(6)	(7)
1. Area under land utilization	25145	25989	25381	25316	25461	25503
2. Forests	1298	2378	1960	1941	2198	2388
	(5.2)	(9.1)	(7.7)	(7.6)	(8 6)	(9.4)
3. Barren and uncultur-able waste	1498	1211	1140	1065	996	790
	(6.0)	(4.7)	(4,5)	(4.2)	(3 9)	(3.1)
4. Land put to non-agricultural use	1852	1845	1912	1988	2034	2162
	(7.4)	(7.1)	(7.5)	(7 9)	(8.0)	(8 5)
5. Culturable waste	2311	1824	1640	1489	1345	1493
	(9.1)	(7.0)	(6.5)	(5.9)	(5.3)	(5.9)
6. Permanent pastures and other grazing land	...	34	44	66	77	279
	(...)	(0.2)	(0.2)	(0.3)	(0 3)	(1.1)
7. Land under miscellaneous tree crops and groves	1240	1018	719	637	615	146
	(4.9)	(3.9)	(2.8)	(2.5)	(2 4)	(0 6)
8. Current fallows	1046	163	145	865	848	940
	(4,2)	(0.6)	(0.6)	(3.4)	(3.3)	(3.7)
9. Other fallow land	291	1281	1260	548	545	605
	(1.2)	(4.9)	(5.0)	(2,2)	(2.1)	(2.4)

(1)	(2)	(3)	(4)	(5)	(6)	(7)
10. Net area sown	15609	16237	16561	16716	16804	16700
	(62.0)	(62.5)	(65.2)	(66.0)	(66.0)	(65.5)
11. Area sown more than once	3650	4070	4463	4651	5550	5615
12. Gross Cropped Area	19259	20307	21024	21368	22353	22378
13. Cropping intensity	123.4	125.1	126.9	127.8	133.0	134.0

*The data do not include information relating to Kumaon and Uttarakhand Division.

Note : Figures in parentheses are percentages to total area under utilization.

Source : Board of Revenue, Uttar Pradesh.

the period of twenty six years and it is an indicator of increase in area sown more than once.

The emerging conclusion from the statistics on land utilisation is that the process of intensive cultivation of land has set in and the percentage share of barren and unculturable waste lands in area under utilization has gone down. It may safely be said that the process of horizontal and vertical expansion of land has taken place simultaneously.

Per capita area sown in Uttar Pradesh

In Table 3.3 below are presented figures of population and net area sown in the State during the last three years of Census:

TABLE 3.3 : **Per capita sown area in Uttar Pradesh**

S. No.	Year	Population (000)	Net Area Sown* (000 hectares)	Per Capita Sown Area (hectares)
	1	2	3	4
1.	1950-51	63216	16231	0.26
2.	1960-61	73746	17188	0.23
3.	1970-71	88341	17305	0.20

* According to agricultural year (July to June)

Source : Column 2, Census of India 1951, 1961 and 1971.
Col. 3 ; Table 3.2

The above table clearly shows that there was an increase of 16.7 and 19.8 per cent in population of the State in the last two decades whereas in the corresponding periods there was an increase of 5.9 and 0.6 per cent respectively in net area sown. The per capita sown area decreased to 0.2 hectares in 1970-71. In fact, the per capita sown area has decreased during the time span of both the decades. The policy implication of decreasing per capita sown area is that the pressure of population on land is increasing and ways and means have to be found out to increase productivity of available land for meeting the growing food needs of the State. It calls for a further breakthrough in farm technology through scientific research.

Weather and crop calendar

The term 'weather' is a vague term. 'Weather' could be taken to mean a collection of various conditions of atmosphere including such phenomena as rainfall, snowfall, hail, humidity, amount of sun-shine, light, atmospheric pressure and temperature.[4] Thus the choice of weather components which might be considered is unlimited. Only rainfall can be used to represent the weather variable as data with regard to other components is not available. Annual rainfall and its month-wise distribution is an important component in the productivity analysis of farm crops. If there is requisite rainfall in the sowing and pre-sowing months pertaining to the crops then their production prospects certainly get brightened. The Tables 3.4 and 3.5 present crop calendar and monthwise distribution of rainfall respectively.

Table 3.4 presents the months of sowing and harvesting of selected farm crops *viz.*, rice, wheat, sugarcane, barley, gram and

TABLE 3.4 : **Sowing and harvesting seasons of selected agricultural crops in Uttar Pradesh**

S. No.	Crop	Sowing	Harvesting
1.	Rice (Autumn)	May-July	September to November
	Rice (Winter)	June-August	November to December
	Rice (Summer)	Jan-February October to December	April to June
2.	Wheat	October to December	March to May
3.	Sugarcane	September to April	October to June
4.	Barley	September to November	March to May
5.	Gram	September to November	March to April
6.	Maize	June-July	August to October

Source : Indian Crop Calendar (Revised and Enlarged Edition) Directorate of Economics and Statistics, Ministry of Food, Agriculture, Community Development and Cooperation, Government of India, New Delhi, 1971.

maize. Intensity of rainfall in the sowing and pre-sowing seasons of these crops can be seen from Table 3.5. In this connection a general inference can be drawn that the distribution of rainfall during the months of the years for all the twentyseven years is quite uneven. Maximum of rainfall is concentrated in the months of June to September. One unique feature of month-wise distribution of rainfall is its erratic nature throughout the period of study. Total annual rainfall ranged between 1414 and 2597 millimetres during 1950-51 to 1976-77. It is a sound proof of uncertain nature of rainfall in the state. It is necessary to mention that too much rainfall i.e., floods as well as too little rainfall i.e., drought should have detrimental influence on productivity of crops grown in the state.[5]

An assessment of the uncertainty and variability of rainfall can be had from the Table 3.6. In this table year-wise dispersion of rainfall in the State over a period of 27 years is presented.

The average rainfall ranged between 141 and 216 millimetres and the year-wise standard deviation ranged between 174 and 322 millimetres during the time period of 27 years. The coefficient of variation ranged between 112.4 and 159.1 per cent. The statistical measures certainly point out that there is maximum variability in month-wise and year-wise rainfall data pertaining to 27 years. Due to uncertain behaviour of rainfall the agriculture of our country has remained a gamble in the hands of monsoon. Extreme dependence on Rain God for purpose of irrigation has been gradually declining with the increase in artificial irrigation facilities.

Growth of irrigation potential in the State

Irrigation has been assigned a crucial role in the strategies adopted for a breakthrough in farm sector of the State because it is the single most important factor which can facilitate the fuller utilisation of scarce farm land resources and can facilitate acceptance of improved farm technology. Irrigation has to be assigned a specific role in agricultural development of the State. Assured irrigation has a protective role against the vagaries of monsoon. Assured water supply can boost the productivity of existing farm inputs even with primitive farm technology. The process of introduction of new farm technology irrigation acts as a catalytic agent

TABLE 3.5 : **Distribution of month-wise rainfall* in Uttar Pradesh, 1950-51 and 1976-77**

S. No.	Year	July	Aug.	Sept.	Oct.	Nov.	Dec.	Jan.	Feb.	March	April	May	June	Annual rainfall (mm)
	(1)	(2)	(3)	(4)	(5)	(6)	(7)	(8)	(9)	(10)	(11)	(12)	(13)	(14)
1.	1950-51	538	831	267	3	0	48	53	23	61	20	25	91	1960
2.	1951-52	312	655	335	13	18	0	23	30	69	5	20	221	1701
3.	1952-53	594	828	178	8	0	13	94	18	5	10	23	224	1995
4.	1953-54	950	721	254	3	0	3	104	117	8	0	15	99	2274
5.	1954-55	571	655	389	117	0	0	89	15	25	8	5	282	2156
6.	1955-56	739	704	439	290	10	0	40	24	44	1	56	233	2580
7.	1956-57	499	789	377	412	56	2	105	31	49	24	17	48	2409
8.	1957-58	598	341	440	27	5	36	22	5	10	16	2	11	1513
9.	1958-59	506	649	672	134	5	24	49	83	17	2	74	65	2280
10.	1959-60	474	666	303	190	31	0	27	2	57	14	12	104	1880
11.	1960-61	789	793	308	337	0	3	96	75	1	10	13	106	2531
12.	1961-62	438	1663	394	295	9	36	89	29	34	6	9	95	2597
13.	1962-63	565	652	578	3	6	23	43	11	23	15	32	202	2153
14.	1963-64	568	851	531	13	15	11	9	7	7	8	71	126	2217
15.	1964-65	751	549	577	21	4	22	17	29	32	60	16	26	2104
16.	1965-66	349	503	324	59	0	2	18	27	3	1	34	370	1690
17.	1966-67	447	572	108	22	24	14	2	3	81	10	12	119	1414
18.	1967-68	672	966	407	10	4	101	56	24	20	6	3	145	2414

19.	1968-69	681	489	189	63	0	6	15	10	21	12	74	98	1658
20.	1969-70	698	750	401	11	37	0	74	88	19	1	68	372	2519
21.	1970-71	460	580	614	98	0	0	32	38	17	83	110	410	2442
22.	1971-72	656	729	400	188	13	0	12	83	9	10	1	86	2187
23.	1972-73	333	477	402	56	36	2	33	22	7	0	55	336	1759
24.	1973-74	540	687	300	208	0	4	1	14	6	1	20	66	1847
25.	1974-75	794	493	150	45	1	27	30	22	22	0	7	386	1977
26.	1975-76	776	571	414	73	0	0	12	22	22	6	9	137	2042
27.	1976-77	539	775	299	15	3	1	31	10	1	20	56	122	1872

The data do not include the rainfall in hill divisions of Kumaon and Uttarakhand Division of the State.

Source (i) Director, Meteorological Centre, New Delhi, and

(ii) Agricultural Situation in India, Directorate of Economics and Statistics, Ministry of Agriculture and Irrigation, New Delhi.

TABLE 3.6 : **Yearwise* dispersion of rainfall in Uttar Pradesh, 1950-51 to 1976-77**

S. N.	Year	Rainfall (mm) Average	Standard Deviation	Coefficient of variation	Index of Total Rainfall
	(1)	(2)	(3)	(4)	(5)
1.	1950-51	163.33	249.97	153 05	100 0
2.	1951-52	141.75	194.38	137.13	86.8
3.	1952-53	189.50	301.46	155.08	116.0
4.	1953-54	189.50	301.46	159.08	116.0
5.	1954-55	179.67	227.50	126.62	110.0
6.	1955-56	215.00	262.28	121.99	131·6
7.	1956-57	200.75	245.03	122.06	122.9
8.	1957-58	126.08	199.95	158.59	77.2
9.	1958-59	190.00	247.23	130.12	116·3
10.	1959-60	156.67	207.71	132.58	95.9
11.	1960-61	210.92	281.71	133.56	129·1
12.	1961-62	216.42	322.11	148.84	132.5
13.	1962-63	179.42	247.87	138.15	109.8
14.	1963-64	184.75	280 01	151 56	113.1
15.	1964-65	175.33	264.11	150.64	107.3
16.	1965-66	140.83	178.97	127.08	86.2
17.	1966-67	117.83	181.26	153.83	72.1
18.	1967-68	201.17	302.80	150.52	123.2
19.	1968-69	138.17	210.63	152.44	84.6
20.	1969-70	209.92	264.54	126.02	128.5
21.	1970-71	203.50	228.67	112.37	124.6
22.	1971-72	182.25	253.86	139.29	111.6
23.	1972-73	146.58	174.20	118.84	89.7
24.	1973-74	153.92	226.85	147.38	94.2
25.	1974-75	164.75	245.62	149.09	100.9
26.	1975-76	170.17	254.54	149.58	104.2
27.	1976-77	156.00	242.94	155.73	95.5

*Rainfall is presented according to agricultural year (July to June).
Source : Table 3.5

and as one of the crucial factors in the package of improved inputs and new technology. A change-over from traditional to modern agriculture can take place with assured irrigation facilities. Reduced extent of fallowing of lands and increased cropping intensity in the State is the result of increased irrigation facilities only.

The Table 3.7 presents the irrigated area by source of irrigation. From the table it is evident that there has been an increase of 65 per cent in the total irrigated area in the State but the relative importance of sources of irrigation has undergone a change during the time period of 26 years. The importance of wells and other sources of irrigation has decreased respectively to the extent of 31 and 12 per cent. Gradually both these sources of irrigation are being replaced by the other sources of irrigation. The increase in irrigated area by tubewells has been to the extent of 1055 per cent. The irrigation by tubewells has become very popular in the State. The area irrigated by canals is increased by 48.5 per cent. Upto the year 1970-71 the irrigated area by canals constituted 34.9 per cent of total irrigated area whereas the corresponding figures for irrigated area by tubewells was 32.5 per cent. Upto 1970-71 the irrigation by canals constituted a major source of irrigation but its importance gradually started decreasing in the current decade and irrigation by tubewells is becoming a major source of irrigation in the State. However, the fact is that the importance of canals as a source of irrigation in the State has not diminished considerably. The emerging conclusions from the data given in Table 3.7 are : (*i*) irrigation by canals is still maintaining its importance as a major source of irrigation in the State, (*ii*) the irrigation by tubewells has grown as a major source of irrigation in the State.

Irrigated area under selected crops

The irrigated area under the selected crops in Uttar Pradesh is presented in Table 3.8. The irrigated area under rice, wheat, sugarcane, and maize increased by 110.7, 200.2, 49.5 and 355.7 per cent respectively whereas the irrigated area under barley and gram decreased respectively by 22.6 and 4.1 per cent. The decrease in irrigated area under barley and gram is partly due to general decrease of land area under these crops. The highest increase was in the irrigated area under maize followed by wheat and rice.

TABLE 3.7 : **Irrigated area* by source of irrigation**

('000 hectares)

S.No	Source of Irrigation	Net Irrigated Area						Percentage increase/ decrease in 1975-76
		1950-51	1955-56	1960-61	1965-66	1970-71	1975-76	
	(1)	(2)	(3)	(4)	(5)	(6)	(7)	(8)
1.	Canals	1847 (38.4)	1736 (35.3)	1993 (39.5)	2292 (39.2)	2498 (34.9)	2743 (34.6)	+48.5
2.	Tube Wells	276 (5.7)	333 (6.8)	543 (10.8)	906 (15.5)	2330 (32.5)	3189 (40.2)	+1055.4
3.	wells	1005 (39.6)	2068 (42.0)	1844 (36.6)	2000 (34.2)	1703 (23.8)	1314 (16 6)	−31.0
4.	Other Sources	781 (16.2)	784 (15.9)	664 (13.2)	647 (11.1)	629 (8.8)	687 (8.7)	−12
5.	Total	4809	4921	5044	5845	7160	7933	+65.0

Source : Board of Revenue, Government of Uttar Pradesh.

Note : Figures in parentheses are percentages to the total irrigated area in each year.

* Information relating to hill portion of Kumaon Division and Garhwal Division is not included in the table.

TABLE 3.8 : **Irrigated area* under selected crops in Uttar Pradesh, 1950-51 to 1975-76**

Year	Rice	Wheat	Sugar-cane	Barley	Gram	Maize
(1)	(2)	(3)	(4)	(5)	(6)	(7)
1950-51	409.0	1627.2	693.1	900·9	402.4	35.7
	(100.0)	(100.0)	(100.0)	(100.0)	(100.0)	(100.0)
1955-56	312.9	1841.7	726.1	920 6	416 4	88.0
	(76.5)	(113.2)	(104.8)	(102.2)	(103.5)	(246.5)
1960-61	501.8	1779.7	874.0	839.5	369.1	68.4
	(122.7)	(109.4)	(126.1)	(93.2)	(91.7)	(191.6)
1965-66	614.0	2223.3	965.0	783.1	546.6	149.9
	(150.1)	(136 6)	(139.2)	(86.9)	(138.8)	(419.9)
1970-71	744.8	3981·8	902.2	736.4	432.9	250.8
	(182.1)	(245.1)	(30.2)	(81.7)	(107.6)	(702.5)
1975-76	861.8	4885.6	1036.1	697.4	385.9	162.7
	(210.7)	(300.2)	(149.5)	(77.4)	(95.9)	(455.7)

* Excluding conventionally estimated area of hill portions of Kumaon and Uttarakhand Divisions.

Note : Figures in parentheses are the indices with 1950-51=100

Source : Directorate of Economics and Statistics, Ministry of Agriculture and Irrigation, Government of India, *Indian Agricultural Statistics*, Volume I, Summary Tables.

Distribution of chemical fertilizers in the State

It is an undisputed fact that the high yielding varieties programme was successful due to assured irrigation and use of chemical fertilizers, besides the use of improved varieties of seeds. In the chemical fertilizers the use of Nitrogen (N_2), Phosphorus Penta Oxide (P_2O_5) and Pottasium Oxide (K_2O) is prevalent in the State. Prior to 1967-68 the use of chemical fertilizers was less prevalent in the State. The use of chemical fertilizers gained momentum during the green revolution. In 1976-77 the index of fertilizer distributed rose to 2514 showing thereby an increase of 2414 per cent in the fertilizer distributed in the State compared to 1950-51. Such a gigantic growth in the chemical fertilizers distributed in the State shows its extent of effectiveness in increasing the productivity per hectare of land. In fact, the use of chemical fertilizer is an important component in vertical expansion of land utilisation programme. For application of scientific methods of cultivation and increasing cropping intensity the role of fertilizers is important. The application of chemical fertilizers to increase fertility of the soil is as important as the application of green manure and compost to preserve its fertility. The chemical fertilizers are useful along with assured irrigation facilities.

The Table 3.9 presents the statistics on distribution of chemical fertilizers in the State for the time period 1960-61 to 1976-77.

It is quite evident that prior to 1966-67 the use of chemical fertilizer was 86 thousand tonnes only. In the year 1970-71 and 1976-77 the chemical fertilizers were distributed to the extent of 411 and 729 thousand tonnes. It is beyond any doubt that the use of chemical fertilizers has been increasing continuously.

Most appropriate would have been to present cropwise consumption of chemical fertilizers in the State. A frantic search for such type of data on yearly basis pertaining to the State did not yield any result. The fertilizers distributed is a rough estimate of the fertilizers consumed in the State.

High yielding varieties programme

Break through in agriculture of the State was brought about by the increased use of chemical fertilizers alongwith the use of high

TABLE 3.9 : **Distribution of fertilizer in Uttar Pradesh 1950-51 to 1976-1977.**

('000 tonnes)

S. No.	Year	N_2O	P_2O_5	K_2O	Total	Index of Fertilizer distributed 1960-61=100
	(1)	(2)	(3)	(4)	(5)	(6)
1.	1960-61	27	2	..	29	100.0
2.	1961-62	24	2	..	26	89.7
3.	1962-63	37	5	..	42	144.8
4.	1963-64	61	8	..	69	237.9
5.	1964-65	89	10	..	99	341.4
6.	1965-66	85	9	..	94	324.1
7.	1966-67	74	9	3	86	296.6
8.	1967-68	142	39	18	199	686.2
9.	1968-69	220	77	42	339	1169.0
10.	1969-70	306	99	55	460	1586.2
11.	1970-71	291	75	45	411	1417.3
12.	1971-72	338	73	53	464	1600.0
13.	1972-73	373	87	52	512	1765.5
14.	1973-74	329	86	48	463	1596.7
15.	1974-75	329	49	32	410	1413.8
16.	1975-76	386	63	38	487	1679.3
17.	1976-77	572	101	56	729	2513.8

Source : Directorate of Agriculture, Uttar Pradesh.

yielding varieties of crops. Jointly both these programmes resulted in the increase of land productivity and increased incomes of the farmers.

The Table 3.10 presents the area under high yielding varieties of rice, wheat, jowar, bajra and maize for the time period 1968-69 to 1976-77. In the year 1967-68 there were 1108 thousand hectares of land area under the high yielding varieties of rice, wheat, jowar, bajra and maize and in the years 1975-76, 1976-77 the area under HYVP increased to 2709 and 7032 thousand hectares respectively. In 1970-71 the area under HYVP of rice, wheat, jower, bajra and maize was 677, 1938, 1.4, 29.3

TABLE 3.10. **Area under HYVP crops in Uttar Pradesh 1968-69 to 1976-77.**

('000 hectares)

S. No.	Year	Paddy	Wheat	Jowar	Bajra	Maize	Total	Index of Total HYVP Area
	(1)	(2)	(3)	(4)	(5)	(6)	(7)	(8)
1.	1968-69	331	1357	8.5	9.8	85	1791.3	100.0
2.	1969-70	561	1639	6.7	22.3	80	2309.0	128.9
3.	1970-71	677	1938	1.4	29.5	63	2708.9	151.2
4.	1971-72	994	2560	0.3	39.7	16	3610.0	201.5
5.	1972-73	1061	3362	0.5	66.5	19	4509.0	251.7
6.	1973-74	1278	3978	1.4	44.8	16	5318.2	296.9
7.	1974-75	1425	4152	0.2	45.0	20	5642.2	315.0
8.	1975-76	1593	4638	0.2	76.8	24	6332.0	353 5
9.	1976-77	1680	5300	2.0	25.0	25	7032.0	392.6

Source : (*i*) Directorate of Agriculture, Uttar Pradesh ; and
(*ii*) Economic Survey of Uttar Pradesh, 1977-78, State Planning Institute, Economics and Statistics Division, Uttar Pradesh, Lucknow.

and 63 thousand hectares respectively and in 1976-77 the corresponding area of these crops increased to 1680, 5300, 2.25 and 25 thousand hectares.

REFERENCES

1. James M. Henderson, *The Utilisation of Agricultural Land. A Theoretical and Empirical Analysis* ; and S.S. Johl and H.S. Sandhu, Existing and suggested cropping patterns by type of forming areas in different districts of Punjab and Haryana, 1969-70, Department of Economics and Sociology Panjab Agriculture University, Ludhiana.
2. Hayami and Ruttan along with their colleagues have discussed with reference to Japan and the Philippines how countries facing land area constraints could still achieve high rates of growth in agriculture through biological innovations and irrigation development, permitting higher productivity on net area sown and increased output through expansion of double cropped area. For details see (i) Hayami and Ruttan, *Agricultural Development : An International Perspectives.* The John Hopkins Press, 1971, (ii) Crisostomo and Barker, *The Philippine Economic Journal*, Growth rates in Philippine Agriculture, 1948-1969, Vol. XI, No. 21, 1972.
3. For details of conventional estimates see Statistical Abstract of Uttar Pradesh, 1973-74, p. 22. State Planning Institute, Economic and Statistics Division, Uttar Pradesh, Lucknow.
4. James L. Stallings, A measure of the influence on weather on crop production, *Journal of Farm Economics*, Vol. 43, December, 1961, pp. 1153-60.
5. (i) Ralph W. Cummings, and S.K. Ray, food grain production relative contribution of weather and new technology, 1968-69 *Economic and Political Weekly*, Sept. 27, 1969 ; (ii) S.K. Ray, Weather and reserve stocks for food grains, *Economic and Political Weekly*, Sept., 25, 1971.

4

Structure and Behaviour of Prices of Selected Agricultural Commodities

SECTION I

Role of price mechanism

IN THIS section a general discussion on the role of price mechanism in an economy is given. The structure and behaviour of prices denote the nature, composition and magnitude of prices. Price mechanism is the medium through which the desires of consumers are transmitted to producers. In fact prices furnish the guide posts to indicate how resources should be utilized in an economy. According to Dr. V.K.R.V. Rao[1] the prime function of price policy is to bring equilibrium between demand and supply of goods. After performing this economic function price movements should be of a self-liquidating nature. In order to avoid a rise in the general price level total output has to be raised to meet the increased demand for individual commodities. The farm prices are the signals which convey the producers demand for agricultural products. Agricultural prices are quickly responsive to demand and supply conditions. Since agricultural output constitutes half of the national product, the general price level is mostly determined by the behaviour of agricultural prices. The fluctuations in the prices of agricultural commodities need be kept to minimum. The producer is to be safeguarded against a sharp price fall following a bumper crop and the interests of the consumers are to be looked after at the time of crop failures or short supplies. For this, price policy should be comprehensive enough to include measures from the production of agricultural products to their distribution. Due to the market imperfections in under-developed countries the effect of price signals gets considerably weakened resulting in either time lags in response or differential magnitudes of responses. Price policy is inevitably a part of the overall economic policy. In a predominantly agricultural country like India the farm prices have a direct bearing on the overall economy. A constant rise in agricultural prices sets an inflationary tendency which may have its serious impact on the economy as a whole. A fall in farm prices is not also in the general interests of the producers since it tends to curb their incentives to produce more.

The classic argument in favour of a mildly rising price level has been provided by Keynes. His preference for a rising price level, though mild, is based on the redistributive effects of a rising price level and on the capacity to create an optimistic atmosphere conducive to growth in investment and output. Commenting on the importance of price policy, Dr. Raj Krishna attributed the three main functions which a price policy has to perform :

(i) to accelerate the growth of agricultural output as a whole,

(ii) to accelerate or decelerate the growth of individual crops or in the context of planning, steer the crop mix according to targets ; and

(iii) to secure adequate increases in the marketed supply of food crops in countries where a large part of output is retained by the peasants for home consumption.[2]

The objective of agricultural price policy is to achieve price stability without destabilising the total revenue of the farmer and provide a price support which would be economic to the farmers as well as to agro-based industries and at the same time safeguard the interests of the consumers. The short term goal of price policy is stability in crop prices to create certainty, the medium term goal emphasises stability in terms of trade for agriculture and the focus of the long term goal is on gradual adjustment of all prices towards their equilibrium level. Thus the role of agricultural price policy is to moderate short run price fluctuations and to allow prices to perform their long run allocative function. In short, it may be emphasized that the aim of the price policy should be to effect a proper structure of production and distribution. Too rigid an insistence on a stable price level may affect the rate of growth adversely and too free a resort to inflation may lead to maldistribution of investment and incomes.

The price mechanism should not be allowed to unduly erode the purchasing power of the various sections of the economy. In India a majority of the population devote a substantial part of their energies to earn incomes sufficient to purchase foodgrains. For this reason, there is no doubt that foodgrain prices should not be among the prices that are to go up on the assumption that foodgrain prices would be stabilised at a remunerative level from the producers' point of view.

Agricultural price analysis is one of the most important and well developed parts of price analysis. The agricultural price analysis is not a mere exercise of academic curiosity but results of it are immediately put to work as bases for huge economic programmes. Agricultural price policy in our country has been under persistent criticism on the ground that it has been deliberately forcing producers' prices down. In 1964, T.W. Schultz[3] asked, "Why are so many poor countries (including India) placing a low economic value on their farm outputs ?". Edward Mason[4] wrote that the prices of foodgrains and some other farm outputs were held down by Government action. Michael Lipton[5] asserted that "farm prices have been systematically kept down since 1960 in India". It is quite true that when the prices of farm products are kept low, the investment incentives for increasing the capacity of agriculture get reduced.

The role of prices in allocation of resources is well known. Factors such as "blending with nature" are to be ignored and it is to be assumed that a farmer employs land and other factors keeping in mind economic objectives. The farmers' main objective under a particular land use is to maximise the economic return from his land. The price which the farmer receives for his product is of vital importance for him. "One of the most permanent aspects of land use decision making is that many events can not be predicted with cent per cent accuracy. Prices at the time of harvest, availability of hired labour, machinery break-down, technological change, government action and weather conditions are examples of factors which effect land use, productivity and income but which are seldom known precisely before they occur."[6] The decision of the farmer to shift the land from one use to another usually takes at least a season. This proposition has been elaborated fully by Dharm Narain as follows :

> In agriculture in particular the biological nature of the production process makes for a considerable lag between production inputs and outputs. The lag between the time of decision in respect of utilisation of land, the most important input factor, and the fruition of that decision into output is still larger.

SECTION II

This section covers the study of behaviour of Farm Harvest prices and Wholesale prices in the State. The time period for the study of behaviour of Farm Harvest prices had to be restricted to 1955-56 to 1975-76 because of non-availability of data for earlier years. The time period in case of Whole Sale Prices extends from 1951-52 to 1975-76.

Price

The farm produce passes from the producer to the consumer after passing through several stages. At the first stage there is the price of a crop received by the producers at the farm or the village site during the period of harvest. At the second stage there is the whole sale price when the farm produce is sold in bulk by the trader. The wholesale price does not relate to any particular season. Thirdly, there is the retail price received by the retail trader and paid by the consumer. This is the price prevalent in the market at any time during the year. The Government of India has been announcing the minimum guaranteed prices known as support prices for the main farm crops. The purpose of such support prices is to save the farmers from the dangers of economically unremunerative prices. The Farm (Harvest) price series is more relevant to the investigation, because due to absence of storage facilities, shortage of required credit needs, compelling needs of debt and interest repayments, a large proportion of farmers sell their farm produce at the time of harvest. In fact it is Farm (Harvest) Price which influences the production and other decisions of the farmers. Such considerations have been kept in view while selecting the relevant price series. However, for making a comparative assessment of the supply response co-efficients the wholesale price indices have also been used.

Farm harvest price data

Since 1948 data on Farm (Harvest) prices have been collected by the Directorate of Economics and Statistics in the Ministry of Agriculture, New Delhi. The data on Farm (Harvest) prices are also being published by the Government of Uttar Pradesh in its Season and Crop Reports. In the year 1947, a uniform definition of Farm (Harvest) price was laid down as follows :

> The farm price may be defined as the average wholesale price at which the commodity is disposed of by the producer at the village site during the specified harvesting period.[8]

The price data have been collected by the government of Uttar Pradesh every week from a number of villages, selected on a purposive basis during the harvest period of six to eight weeks. Then the weekly prices were combined into tehsil and district averages through their simple arithmetic mean. The State average was calculated as a weighted average of district prices with the production of crop in the district as weights. Price quotations were collected through the revenue agencies operating in the State.

Wholesale price indices*

The second price series used in the investigation is the Wholesale Price Series which was published by Economics and Statistics Division, State Planning Institute, Lucknow. Analysis of wholesale price indices was available on the base 1957-58=100 and for the sake of maintaining uniformity the farm harvest price indices were also constructed on the base 1957-58=100.

Movements of farm harvest prices of rice, wheat and sugarcane

Analytical study of the movements of the Farm Harvest prices

* Details of computations are given in Appendix 4.1.

of the selected farm commodities *viz.*, rice, wheat, sugarcane, barley, gram and maize has been made in this section. The farm harvest prices along with their index numbers relating to rice, wheat and sugarcane are presented in Table 4.1.

Paddy

The price of paddy is presented in col. (2) of Table 4.1 while the index numbers of these prices with 1957-58 as base year are presented in col. (3). It is seen from the table that the price of

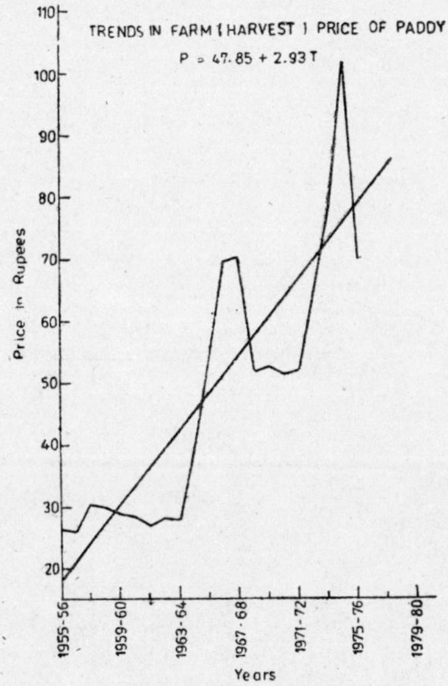

GRAPH 4A

paddy fluctuated between Rs. 26.04 and Rs. 101.71 per quintal. The index of price of paddy was highest in the year 1974-75 when it stood at 335.7. The year 1974-75 recorded an all time high record price of Rs. 101.71 per quintal. In the year 1975-76 there was a general decline in the price of farm commodities and the price of paddy also came down to Rs. 69.32 per quintal. Excluding the year 1974-75 the price of paddy ranged between Rs. 26.04 and Rs. 77.27 per quintal. During the period 1955-56 to 1963-64 the price of paddy oscillated around Rs. 30 per quintal. From the year 1964-65 onwards there was a continuous rise in the price of paddy and in the year 1967-68 the price rose to Rs. 70.23 per

TABLE 4.1 : **Price indices of farm (harvest) price of paddy, wheat and sugarcane**

(Price in Rupees per Quintal)

	Year/Crop	Paddy		Wheat		Sugarcane	
		Price	Index	Price	Index	Price	Index
	(1)	(2)	(3)	(4)	(5)	(6)	(7)
1.	1955-56	26 04*	85.9	35 26	85.50	3.59	98.6
2.	1956-57	26.04	85.9	39.97	96.90	3.62	99 5
3.	1957-58	30.30	100.0	41.26	100.0	3.64	100.0
4.	1958-59	29.95	98.8	44.58	108.1	3.62	99.5
5.	1959-60	28.99	95.7	42 65	103 4	3.62	99.5
6.	1960-61	28.53	94 2	39.73	96.3	4 05	111 3
7.	1961-62	26.85	88.6	39.60	96.0	3.80	104.4
8.	1962-63	28.18	93.0	39.97	96.9	3.86	106 0
9.	1963-64	27.78	91.7	54,91	133.1	5 09	139.8
10.	1964-65	40.05	132.2	71.20	172.6	5.27	144.8
11.	1965-66	52.71	174.0	70.44	170.7	5.03	138.2
12.	1966-67	69.38	229·0	116.35	282.0	5.45	149.7
13.	1967-68	70,23	231.8	78.27	189,7	13,67	375.6

14.	1968-69	51.73	170.7	79.00	191.5	9.42	258.8
15.	1969-70	52.80	174.3	88.56	214.6	6.82	187.4
16.	1970-71	51.64	170.4	75.31	182.5	6.83	187.6
17.	1971-72	51.85	171.1	76.13	184.5	8.28	227.5
18.	1972-73	63.42	209.3	82.23	199.3	11.60	318.7
19.	1973-74	77.27	255.0	137.91	334.3	11.80	324.2
20.	1974-75	101.71	335.7	143.26	347.2	12.74	350.0
21.	1975-76	69.32	228.8	98.47	238.7	10.13	278 3

* Price of Paddy for the year 1955-56 has been assumed to be the same as prevailed in the year 1956-57.

Sources :

1. Cols. (2) & (4), Directorate of Economics and Statistics, Ministry of Agriculture and Irrigation, Krishi Bhavan, New Delhi.
2. Col. (6), State Planning Institute, Department of Economics and Statistics, Lucknow, Uttar Pradesh : Vide Letter No ; 967/Sub.-42/76 dated 27th May, 1978.

quintal. During the period 1968-69 to 1971-72 price stability was noticed and the price stood roughly at Rs. 52 per quintal during this period.

During the entire period under investigation the mean price level stood at Rs. 47.85 per quintal with Rs. 20.84 standard deviation and 43.55 per cent coefficient of variation. On excluding the year 1974-75 from the analysis the coefficient of variation worked out to be 38. 65 per cent, thereby showing less variation in price compared to the series including the year 1974-75.

The graph 4A depicts the trends in the price of paddy during 1955-56 to 1975-76. The trend values were obtained by fitting linear regression model

$$P=A+BT$$

P	Price
T	Time
A	Constant
B	Rate of growth

The regression equation obtained on the basis of price data of paddy is $P=47.85+2.93\ T$.

Wheat

The price and indices of price of wheat are respectively presented in columns (4) and (5) of Table 4.1. Excluding the year 1963-64 the behavioural pattern of price of wheat roughly resembled the price behaviour of paddy during the entire period of investigation. The price of wheat ranged between Rs. 35.26 and Rs. 143.26 per quintal. Excluding the year 1958-59 the price of wheat oscillated around Rs. 40 per quintal during 1955-56 to 1962-63. During 1962-63 and 1966-67 the price continued to rise and the price of wheat rose to the level of Rs. 116.35 per quintal in 1966-67. The highest price of wheat was noticed in 1974-75 when the price index rose to 347.2 compared to 1957-58.

During the entire period the mean level of price worked out to Rs. 71.19 per quintal. The standard deviation and coefficient of variation of the price series of wheat worked out to Rs. 74.55 and Rs. 104.72 per cent respectively. The price fluctuations in case of wheat were more pronounced compared to the price of paddy.

The graph 4B depicts the trends in price of wheat obtained on the basis of linear regression equation

$$P=A+BT$$

where

P	Price of wheat
A,B,T	have usual meaning.

The regression equation obtained on the basis of price data of wheat is $P=71.19+4.39\ T$.

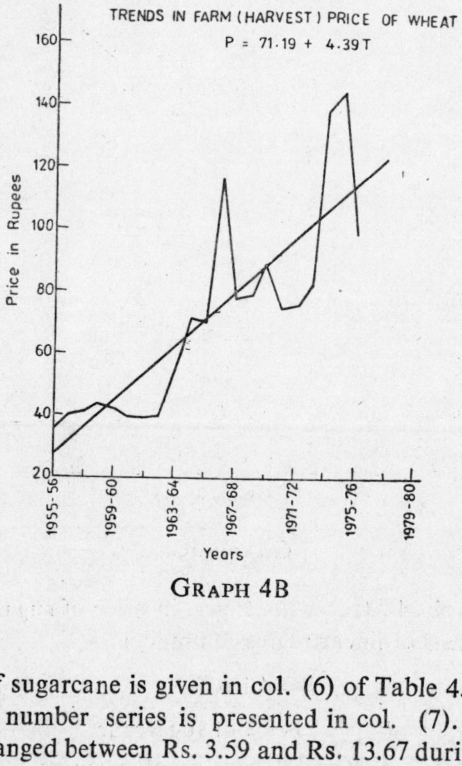

GRAPH 4B

Sugarcane

The price of sugarcane is given in col. (6) of Table 4.1 whereas its price index number series is presented in col. (7). The price of sugarcane ranged between Rs. 3.59 and Rs. 13.67 during the period of investigation. The price of sugarcane reached the highest record level of Rs. 13.67 per quintal in the year 1967-68 followed by Rs. 12.74 per quintal in 1974-75. During the period 1955-56 to 1966-67, the price of sugarcane oscillated in the range of Rs. 3.0 to Rs. 5.45 per quintal. The price of sugarcane fell to Rs. 10.13 per quintal in 1975-76.

The average price of sugarcane worked out to be Rs. 6.76 per quintal. The standard deviation and co-efficient of variation of the price series worked out to be Rs. 3.35 and 49.56 per cent respectively. The variation in price of sugarcane has been greater than the price variation in paddy whereas this price variation is lesser than the price variation observed in wheat.

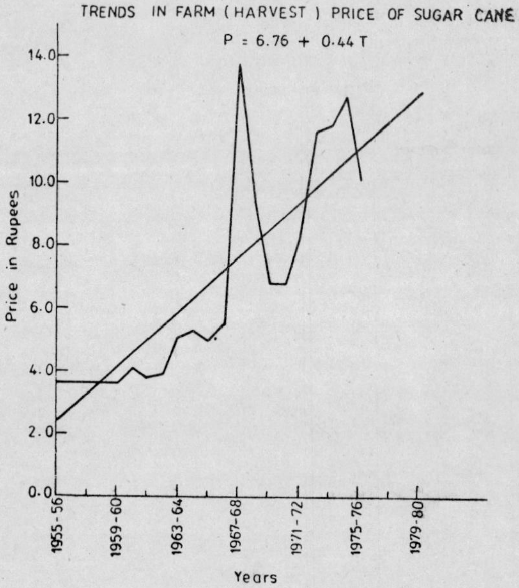

GRAPH 4C

The graph 4C traces the trends in price of sugarcane worked out on the basis of linear regression model

$$P=A+BT$$

P Price of sugarcane

A, B and T have usual meaning.

The regression equation obtained on the basis of price data of sugarcane is $P=6.76+0.44\ T$.

Movements of farm harvest prices of barley, gram and maize

The price and price index numbers pertaining to barley, gram and maize are presented in Table 4.2.

TABLE 4.2 : **Price indices of farm (Harvest) price of barley, gram and maize**

(*Price in rupees per quintal*

	Year/Crop	Barley		Gram		Maize	
		Price	Index	Price	Index	Price	Index
	(1)	(2)	(3)	(4)	(5)	(6)	(7)
1.	1955-56	25.67	88.4	26.31	91.4	28.64*	96.2
2.	1956-57	31.16	107.3	32.23	112.0	28.64	96.2
3.	1957-58	29.04	100.0	28.77	100.0	29.77	100.0
4.	1958-59	34.51	118.8	36.46	126.7	36.52	122.7
5.	1959-60	31.75	109.3	32.23	112.0	29.45	98.9
6.	1960-61	31.08	107.0	33.41	116.1	31.78	106.8
7.	1961-62	31.03	106.9	34.29	119.2	29.50	99.1
8.	1962-63	28.10	96.8	36.04	125.3	29.58	99.4
9.	1963-64	44.07	151.8	51.05	177.4	31.72	106 6
10.	1964-65	53.24	183.3	57.87	201.1	49.89	167.6
11.	1965-66	57.47	197.9	63.31	220.1	56.07	188.3
12.	1966-67	90.55	311.8	110.75	384.9	63.38	212 9
13.	1967-68	60.37	207.9	76.78	266.9	82.91	278.5

	(1)	(2)	(3)	(4)	(5)	(6)	(7)
14.	1968-69	55.57	191.4	74.59	259 3	53.01	178.1
15.	1969-70	61.69	212.4	87.25	303.3	56 74	190.6
16.	1970-71	51.51	177.4	69.00	239 8	47.68	160.2
17.	1971-72	59.27	204.1	85.27	296.4	51.33	172.4
18.	1972-73	81.24	279.8	117.19	407.3	68 74	230.9
19.	1973-74	118 18	407.0	191.63	666.1	76.51	257 0
20.	1974-75	119 20	410.5	172.62	600.0	153.00	513.9
21.	1975-76	61.01	210.1	122 27	425.0	68.49	230 1

*Price of Maize for the year 1955-56 has been assumed to be the same as prevailed in the year 1956-57.

Source : 1. Cols (2), (4) and (6), Directorate of Economics and Statistics, Ministry of Agriculture and Irrigation, Krishi Bhavan, New Delhi.

Barley

The price and price index numbers (1957-58=100) of barley are presented in columns (2) and (3) of Table 4.2. From the table it can be seen that the price of barley ranged between Rs. 25.67 and Rs. 119.20 per quintal during the entire period of investigation. During the period 1958-59 to 1962-63 the price of barley continuously decreased year after year. During the period 1962-63 to 1966-67 the price continuously increased year after year. During the period 1967-68 to 1970-71 the price increased in every alternate year. In the years 1973-74 and 1974-75 the price of barley was respectively Rs. 118.18 and Rs. 119.20 per quintal. During these two years the price of barley was quite high compared to the price which prevailed in the other years of the series. In the year 1975-76 the price of barley decreased to the level of about 50 per cent of the price which prevailed in 1974-75. From the series of index numbers of prices it is noticed that the highest price index rose to 410.5 showing an increase of 310.5 per cent in the year 1974-75 compared to the base year 1957-58.

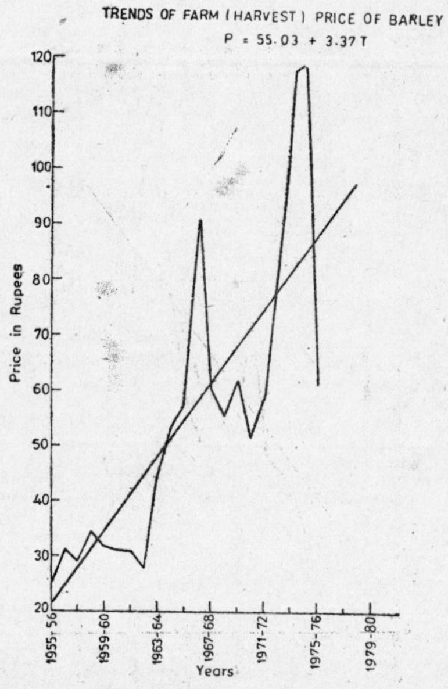

GRAPH 4D

The average price and standard deviation of the price series of barley worked out to Rs. 55.03 and Rs. 26.97 respectively. The coefficient of variation of the price series worked out to be 49.01 per cent.

The graph 4D traces the trends in the price of barley worked out on the basis of linear regression equation

$P=A+BT$

P Price of barley

T Time

A and B having the usual meaning.

The regression equation obtained on the basis of price data of barley is $P=55.03+3.37\ T$.

Gram

The price and price index numbers (1957-58=100) of gram are presented in the columns (4) and (5) of table 4.2. It can be seen from the table that the price of gram ranged between Rs, 26.31 and Rs. 191.63. The price of gram increased by 566.1 per cent during the period 1957-58 to 1973-74. Such an alarming rate of

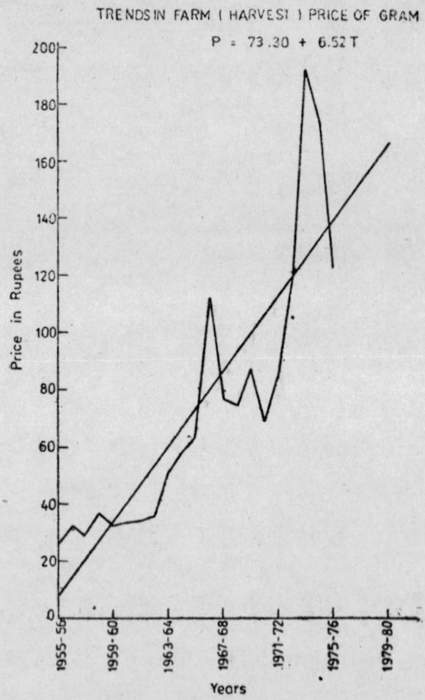

GRAPH 4E

increase was not noticed in the price of other selected crops. During the period 1959-60 to 1966-67 the price of gram increased continuously year after year. In the year 1966-67 the price of gram touched the record level of Rs. 110.75 per quintal. The rate of increase in price during the period mentioned above was to the extent of 284.9 per cent. In the year 1975-76 the price of gram decreased to Rs. 122.27 per quintal and the price index fell to 425 from 600 in 1974-75.

The average price and standard deviation worked out to Rs. 73.30 and Rs. 83.16 respectively. The coefficient of variation of the price series worked out to 113.45 per cent.

The graph 4E traces the trends in the price of gram obtained on the basis of regression model

$$P=A+BT$$

where

P Price of gram

A, B and T have usual meaning.

The regression equation obtained on the basis of price series of gram is $P=73.30+6.52T$.

Maize

The price and price index numbers (1957-58=100) are presented in columns (6) and (7) of table 4.2. The price of maize ranged between Rs. 28.64 and Rs. 153.00 per quintal. If the year 1974-75 is excluded from the series the price range falls between Rs. 28.64 and Rs. 82 91 per quintal. These are two main trends discernible in the price series. First, the price increased continuously during the two periods *viz.*, 1961-62 to 1967-68 and 1970-71 to 1974-75, Second, in the rest of the period of investigation no such a trend was observed. In the year 1975-76 the price of maize decreased to Rs. 68.49 per quintal and the index number of price of maize fell to the level of 230.1. The highest price index of 513.9 was noticed in 1974-75.

The average price and standard deviation of the price series of maize worked out to Rs. 52.54 and Rs. 28.17 respectively. The coefficient of variation of the price series worked out to 53.62 per cent.

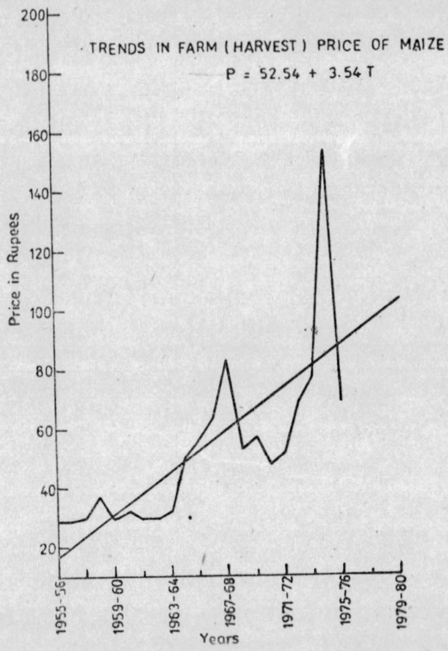

GRAPH 4F

The graph 4F depicts the trends in the price of maize. The trend values have been obtained by fitting the regression model

$P = A + BT$ to price data of maize.

P Price of maize

A, B and T have usual meaning.

The regression equation obtained on the basis of price data of maize is $P = 52.54 + 3.54T$.

Growth rates at a glance

A summary of linear growth rates of farm harvest price series of the selected commodities is presented in Table 4.3. If the period of investigation is considered as a whole, a definite rising trend of farm harvest prices is visible in case of all the selected crops. Last year of the series is exceptional where a notable decline is displayed in prices of all commodities. The percentage increase of prices in 1975-76 over 1955-56 was highest in case of gram (364.73 %) followed by sugarcane (182.17 %) and the minimum increase was 137.67 per cent in barley. For the superior cereals

TABLE 4.3 : **Farm harvest prices of selected crops in 1955-56 and 1975-76**

(Price in rupees per quintal)

Crop	Farm Harvest Price 1955-56	Farm Harvest Price 1975-76	Percentage over 1955-56	Linear Rate of Growth	Coefficient of variation of price series 1955-56 to 1975-76
(1)	(2)	(3)	(4)	(5)	(6)
1. Paddy	26.04	69.32	166.21	2.93	43.55%
2. Wheat	35.26	98.47	179.27	4.39	104.72%
3. Sugar-cane	3.59	10.13	182.17	0 44	49.56%
4. Barley	25.67	61.01	137.67	3.37	49.01%
5. Gram	26.31	122.27	364.73	6.52	113.45%
6. Maize	28.64	68.49	139.14	3.54	53.62%

Source : Computed from Tables 4.1 and 4.2.

like rice and wheat the price increase was 166.21 per cent and 179.27 per cent respectively. In case of maize the price increase in farm harvest in 1975-76 over 1955-56 was to the extent of 139.14 per cent.

The linear growth rates of farm harvest prices of gram was highest 6.52 per cent per annum, followed by wheat 4.39 per cent and the lowest linear growth rate was 0.44 per cent per annum for sugarcane preceded by 2.93 per cent per annum in case of paddy. The increase in price per annum in case of maize and barley was 3.54 per cent and 3.37 per cent respectively.

Movements of wholesale prices of selected farm commodities

In order to analyse the behaviour of price movements of the selected crops during the period 1951-52 to 1975-76, the price indices taking 1957-58 as base presented in Table 4.4 may be observed. An analysis of wholesale price indices of the selected farm commodities is presented below :

Rice and wheat

During the period prior to base year 1957-58 the price indices of both rice and wheat were lower during the period 1953-54 to 1956-57 and were higher during 1951-52 to 1952-53 compared to the base year. During 1957-58 to 1962-63 there was a consistent

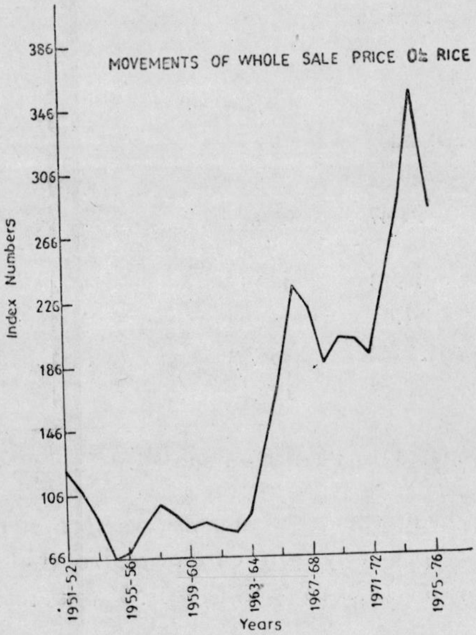

GRAPH 4G

fall in price and the price indices of rice and wheat came down by 18.2 and 6.1 points respectively. The fall in price of rice was remarkably more compared to the price of wheat. During 1962-63 to 1966-67 there was continuous rice in the whole sale price index of rice and its price index rose by 135.2 points. The price indices of wheat showed a haphazard pattern during the period 1962-63 to 1971-72. During 1971-72 to 1974-75 a very steep rise in the price indices of both rice and wheat was observed. In case of rice the index rose to 354.3 compared to the index 414.4 in case of wheat in the year 1974-75. The last year *viz.*, 1975-76 was a year of general price decline and the price index of rice and wheat came down to 281.8, 248.5 points respectively. The graphs

TABLE 4.4. : **Wholesale price indices of selected crops, 1951-52 to 1975-76**

	Year	Rice	Wheat	Gur	Barley	Gram	Maize
	(1)	(2)	(3)	(4)	(5)	(6)	(7)
1.	1951-52	120 2	103.8	101.2	106 1	117.3	105.4
2.	1952-53	108.6	117.5	89.1	110.4	131.0	102.3
3.	1953-54	90.7	99.9	115.7	97 8	111.0	90.7
4.	1954 55	66.0	78.4	97.1	64 9	64.0	60.3
5.	1955-56	69.7	83.3	79.4	80.0	80 6	74.2
6.	1956-57	86 0	99.9	87 0	107.9	110.9	100.2
7.	1957-58	100.0	100 0	100.0	100.0	100.0	100 0
8.	1958-59	93 8	134.9	123.3	140.2	148.8	117.1
9.	1959-60	85.3	108.3	115 3	107.3	109.0	91.2
10.	1960-61	87 9	104.4	100.8	108.6	127.6	88.7
11.	1961-62	83.7	98.5	106.7	98.5	121.1	89.2
12.	1962-63	81.8	93 9	140.2	90.2	128.4	88.4
13.	1963-64	91.8	122.3	171.4	129.3	159.2	108.8
14.	1964-65	136 0	196.7	172.4	210.3	230.7	166.9
15.	1965-66	179.3	185.6	151.4	206 6	224.6	183.4
16.	1966-67	235.2	246.1	240.2	266.2	323.6	279.2

	(1)	(2)	(3)	(4)	(5)	(6)	(7)
17.	1967-68	221.2	245 1	447.9	268 5	367.8	264.1
18.	1968-69	187.5	189.5	318.5	176.9	260.4	173 9
19.	1969-70	202.4	217.3	169.8	218.7	334 4	201.9
20.	1970-71	200.5	184.6	189.2	158.8	251.0	169.8
21.	1971-72	192.1	191.4	292.6	183.5	294.3	207.3
22.	1972-73	240.0	227.8	383.4	266 6	396.7	223.6
23.	1973-74	284.8	291.7	350.2	377.5	611.1	310.3
24.	1974-75	354.3	414.4	386.6	439.0	660.0	437.0
25.	1975-76	281.8	248.5	404.0	228.3	496.9	273.2

Source ; Prices in Uttar Pradesh (1950-1975) : Part III (Price Indices) : Economics and Statistics Division, State Planning Institute, Uttar Pradesh, February, 1978.

4G and 4G′ depict fully the price movements of both the commodities during the period 1951-52 to 1975-76.

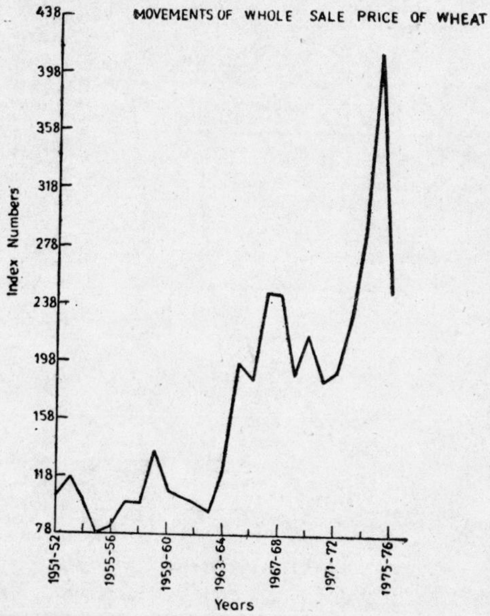

GRAPH 4G′

*Gur**

During the period prior to the base year 1957-58 the price indices of gur were lower except for the years 1951-52 and 1952-53. During the time periods 1960-61 to 1964-65 and 1969-70 to 1975-76 the price indices of gur rose continuously. In the rest of the years the price indices did not show any specific pattern. It is remarkable to note that indices of all the selected commodities except the price Index of gur fell in the year 1975-76, a year of general price decline. The graph 4H exhibits the behaviour of price of gur during the period 1951-52 to 1795-76.

Barley, Gram and Maize

In case of all the three farm commodities *viz.*, barley, gram and maize a common feature of price rise was observed during the period 1970-71 to 1974-75 compared to base year 1957-58. In

*** Gur is a kind of raw sugar manufactured by indigenous methods.**

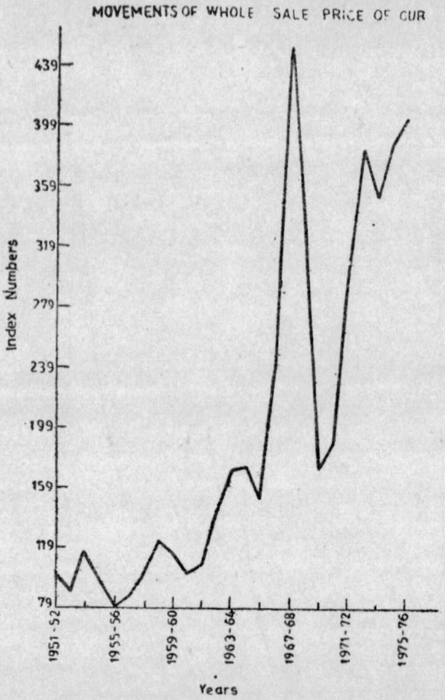

GRAPH 4H

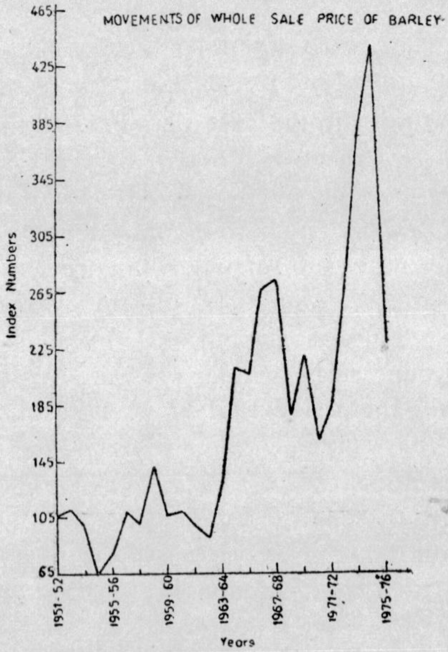

GRAPH 4I

1975-76 a decline in the price index of all the three farm commodities was observed. The wholesale price index number of barley fluctuated between 80.0 and 218.7 whereas the wholesale price index numbers of gram and maize fluctuated between 64.0 and 367.8 ; and 60.3 and 201.9 respectively. The graphs 4I, 4I′ and 4I″ depict movements of wholesale prices of barley, gram and maize respectively.

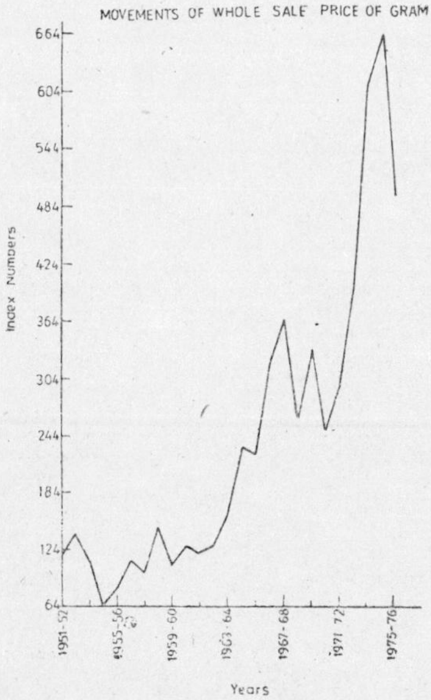

GRAPH 4I′

From the analysis of price behaviour of the selected crops it is seen that the movements of farm harvest price and wholesale price pertaining to the selected commodities were roughly in the same direction.

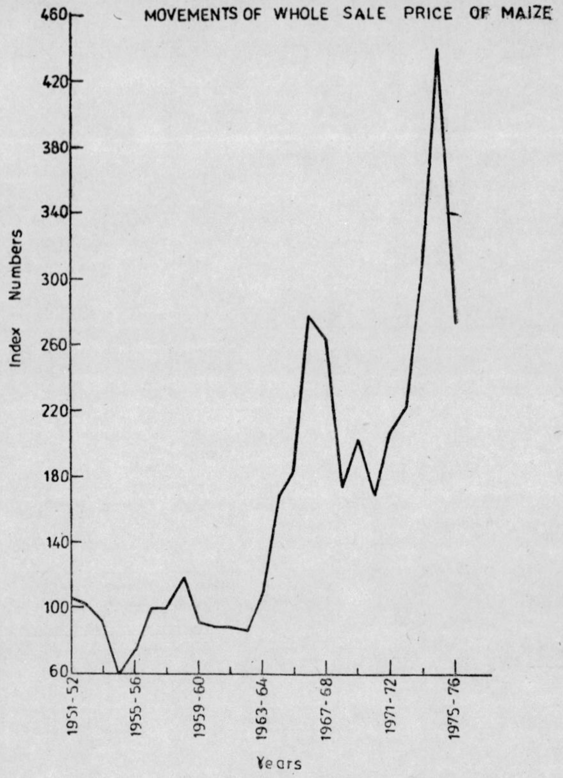

GRAPH 4I"

REFERENCES

1. Dr. V.K. R.V. Rao, *Essays in Economic Development*, Chapter 6.
2. Raj Krishna, Agricultural price policy and economic development in *Agricultural Development and Economic Growth*, edited by H.S. South Worth and Bruce F. Johnston, Cornell University Press, New York, 1967, p. 503.
3. T.W. Schultz, Production opportunities in Asian Agriculture ; An Economist's Agenda, Agricultural Economics Paper No. 68 : 12, Department of Economics, University of Chicago (Revised July 12, 1968).
4. Edward Mason, *Economic Development of India and Pakistan*, Harvard University Press, Cambridge, 1966.
5. Michael Lipton, Urban bias and food policy in poor countries, *Food Policy*, Vol. I, November, 1975.
6. Found William C., *A Theoretical Approach to Rural Land-Use Patterns*, Edward Arnold (Publishers) Ltd., London, 1971.
7. Dharm Narain, *Impact of Price Movements on Areas Under Selected Crops in India*, 1900-39, Cambridge University Press, London, 1965.
8. Farm Prices of Principal Crops, 1947-48 to 1951-52, Ministry of Food and Agriculture, Government of India, New Delhi.

5

Supply Response of Rice, Wheat and Sugarcane

THIS chapter deals with (*i*) survey of relevant literature on Supply Response, (*ii*) methodological issues pertaining to variables used in *Nerlovian Distributed Lag Model* and (*iii*) empirical estimation of the coefficients of supply response models pertaining to the crops of rice, wheat and sugarcane.

On the demand side, a rapid growth of population accompanied by a rise in income levels leads to an increase in demand for agricultural and non-agricultural commodities. On the supply side, the economic activities are governed by the motive of profit in both the sectors and production decisions are guided by price mechanism. Price mechanism is supposed to lead to economic decisions based on allocative efficiency. The price system is a powerful tool to elicit and transmit essential economic information to stimulate appropriate decisions by producers and consumers.[1] Thus an important aspect of the problem is to study the response of supply of agricultural products to changes in prices.

Social and natural scientists, especially geographers contend that differences in land-use can be explained to a considerable extent by differences in quality of land. This ecological relationship is so important and overwhelming that the role of farmer as a decision maker can even be disregarded.[2] This is quite true that ecological relationships may limit the range of choices, but some choices viz., the use of same land for different crops either in combination or as a substitute for one another, continue to exist for the farmers. Further, the quality and quantity of inputs other than land may be varied by producers even after a particular crop has been sown on the farm. Generally the factors such as 'blending with nature' are ignored by economists and it is assumed that a farmer employs land and other factors with economic objectives only.

The importance of price response of supply is derived from the fact that a perverse relationship or absence of relationship between changes in prices and supply blunts the policy instrument for action of government of a country. The comments of Messrs Galbraith and Black on the effectiveness of price mechanism in the farm sector are worth consideration. "For most individual farm products the evidence is that their output is highly responsive to price if a little time is allowed for adjustment. But if the response is that of the total output of a given group of farms then indeed it is slow."[3] But the same author adds ;

> There is not a little evidence to show that, of the outputs which are technically possible, the farmer will produce a collection very much like that which would yield the greatest profit, although, in fact, he may never operate at the point of equilibrium.[4]

Marc Nerlove[5] developed a methodology for finding the impact of price changes on the acreage allocation among different crops. The methodology developed by him is popularly known as *Nerlovian Distributed Lag Adjustment Model.* In the field of agricultural economics this model has been generously used by research workers. However, some doubts were cast on the applicability of the model to underdeveloped countries because, in these countries the production decisions of the farmer were considered to be dominated by considerations relating to soil-climatic conditions, consumption requirements of the family and cash needs. With reference to these countries the significant influence of general market forces reflected through changes in relative prices was not recognised.

It is quite true that in underdeveloped countries a significant proportion of farm land is devoted to the production of domestic requirements by the farmers, but in sowing the surplus agricultural land the farmers have to look to the market. Therefore, the impact of price on acreage allocation seems to be quite evident.

A number of researchers were prompted to carry out the investigations for testing the farmers' responsiveness to price variation in underdeveloped countries. In India also, a number of studies were carried out in post planned period for testing the farmers' responsiveness to price variation in allocation of farm land among different crops. A survey of research in economics contains a detailed information pertaining to research done in India in the field of supply response.[6] These studies range from investigation into association between changes in prices and changes in area under specific crops, to formulation and testing of dynamic models of supply function. Pioneering studies of Dharm Narain and Raj Krishna need a special mention here. Dharm Narain's study has significance of its own.[7] The gigantic effort made by him in obtaining the data and the height of imagination in coordinating and filling the gaps in the data is commendable. In the study the author has investigated the association between changes in prices and areas under different crops. The tabular presentation and geometric charts bring into clear focus the positive relationship between acreage of competing crops and relative prices. The relationship of output, area with changes in weather and prices have been also indicated in the study. The investigation revealed that between food and non-food crops, the production of former is influenced to a greater extent by weather and that of the latter by prices. An element of subsistence economy is believed to be responsible for the differential behaviour. The author has employed graphic methods for analyzing the complex economic relationships. Although such analysis has the merit of simplicity yet these lack mathematical precision. In this study no econometric approach has been attempted and therefore, it has not been possible to quantify the influence of different relevant variables.

Raj Krishna used Nerlovian Distributed Lag Model in his investigation.[8] It is maintained that due to various types of uncertainties in agriculture the response of farmers to adjust supplies to changes in prices may be slow and spread over a period of time and level of production achieved may be different from the planned level. It has been assumed in the study that instead of current prices, the prices prevailing in the preceding year would influence the planned level of production. The investigation revealed that the farmers in the erstwhile Punjab adjusted acreage under competing crops like wheat and cotton in the same manner as the farmers in the United States of America do. In other words, economic forces operate both in the under-developed and highly developed countries in the same manner. This study ruled out the need for explaining the behaviour of the so-called subsistence farmer in terms of non-economic factors.

G. Parthasarthy[9] and P.V. John[10] in their studies observed a negative relationship between supply and prices. Both the authors argued that the relative profitability of sugarcane rather than its price was the chief factor determining the change in acreage.

S.C. Gupta and A. Majid[11] in their study found the relative profitability of sugarcane, among other factors, to be responsible for the changes in acreage. Chandresh Kumar[12] in his study reinforced the conclusions of Raj Krishna and Dharm Narain. Y. Satyanarayan[13] in his study observed that the installed capacity of sugar factories had a positive linear relationship with sugarcane acreage. It is clear that besides prices there are some non-price variables also which influence the acreage alloca-tion among different crops.

J.R. Behrman[14] analysed the responsiveness of four major annual crops : rice, cassava, corn and kenaf in Thailand. In his study no significant evidence has been found for the hypothesis that farmers in underdeveloped countries are more influenced by the institutional constraints prevailing in the economy. The study concludes that farmers in underdeveloped countries respond significantly to price changes. Hossein Askari and Johan Thomas Cummings[15] have presented details of research done in the field of supply response.

Keeping in view the numerous studies already completed in the field of supply response it is quite clear that the variations present in the results obtained in such studies pertaining to underdeveloped countries leave a great scope for further research in this field. Keeping it in view the present study with respect to an underdeveloped region of Uttar Pradesh is undertaken. The results obtained with reference to Uttar Pradesh in selected supply response studies will be compared with the results obtained in the present study.

METHODOLOGICAL ISSUES AND THE VARIABLES SELECTED FOR NERLOVIAN MODEL

The model used by Raj Krishna is Nerlovian 'adjustment' model :

(1) $X_t^* = a + bP_{t-1} + CY_{t-1} + gZ_{t-1} + h_{wt} + u_t$

(2) $X_t - X_{t-1} = B(X_t^* - X_{t-1})$

where

X_t^* is the standard irrigated acreage that farmers would plant in the year t if there were no difficulties of adjustment, X_t is the standard irrigated acreage actually planted to the crop in the harvest year t. The 'standard irrigated' acreage of the crop is the irrigated acreage plus the unirrigated acreage multiplied by a standardisation factor.

P is the relative price of the crop, i.e. the post-harvest price of the crop deflated by an index of the post-harvest prices of the alternative crops.

Y is the relative yield of the crop, *i.e.*, the yield of the crop deflated by an index of the yields of alternative crops.

Z is the total irrigated area in all crops of the season

W is rainfall

u is stochastic disturbance term.

B is the Nerlovian coefficient of 'adjustment'.

The farmers, it is postulated, are able to increase the acreage of a crop in any year only to the extent of a fraction *B* of the difference between the acreage they would like to plant and the acreage actually planted in the preceding year.

Equations (1) and (2) yield the estimating equation ;

(3) $X_t = a_0 + b_2 P_{t-1} + b_3 Y_{t-1} + b_4 Z_{t-1} + b_5 W_t + b_6 X_{t-1} + V_t$

where

$a_0 = aB$, $b_2 = bH$, $b_3 = cB$, $b_4 = gB$, $b_5 = hB$, $b_6 = (1 - B)$ and $V_t = B^u{}_t$

Out of the 'shifter' variables *Y*, *Z* and *W*, shifting the acreage price relation, only those have been included in the equation of a given crop by the author which were found, on a preliminary analysis, to be important factors determining the acreage of that particular crop.

METHODOLOGICAL ISSUES

Comments no variables used in Nerlovian 'Adjustment' models

Acreage as dependent variable

Acreage under the three selected crops *viz.*, rice, wheat and sugarcane has been taken as the dependent variable. The reason for using acreage as dependent variable is that Nerlovian Model is not directly estimable because of unobservable variable x_t^* and the time series estimates of planned output are not available. Actual output could be one possible proxy. Realized agricultural output often differs considerably from planned output because of many environmental factors beyond the control of the farmer. It is the discrepancy between planned and realized output in any production period that led to approximate planned output, to which the supply function actually refers, by acreage planted. Therefore, the area planted under a particular crop is considered as dependent variable.

It is quite necessary to point out the inadequacies of approximation of planned production by planted area. First, land is only one of many inputs in agriculture. It will thus be preferable to use an index of all inputs for the crop. However, most land inputs are not committed to one crop only, but are changed throughout the crop period in response to factors over which the farmer has no control. Also, it is very difficult to have data on the quantity of such factors used for a specific crop. And non-availability of data precludes the construction of an all inclusive index. Secondly, land is not a homogeneous factor, rather it is a heterogeneous factor. If there is a constraint of other inputs, a farmer can increase his planned output by using better, though less land. Nerlovian model assumes away this heterogeneity by taking correlation coefficient between planned output and planted area near to one[16].

Relative price as independent variable

The change in acreage allocation in response to changes in price takes place with a time lag. This reasoning has led to gauge the impact of lagged values of price variable on the acreage planted. Ordinarily the price variables has been used with a lag of one year and in selected situations the price has been used with a lag of two years. The study of impact of relative price changes in shaping the farmer's decision in acreage allocation among selected crops is the crux of the study. The examination of the effect of price on acreage changes under a particular crop with reference to that crop only is inadequate. The prices relevant to the farmer in sowing decisions are the prices of crops which compete for area with it. When a number of crops compete for the same area then the price of a particular crop needs to be corrected for the changes in price of competing crops. Keeping it in view the price of rice has been corrected for changes in prices of competing crops of sugarcane and maize, the price of wheat has been corrected for changes in prices of competing crops of sugarcane, barley and gram ; the price of sugarcane has been corrected for changes in prices of competing crops of rice and wheat. The composite index has been used for applying such corrections. The composite index has been constructed by assigning weight to price indices of the competing crops according to the area these crops occupied at the average of initial three years. The weights so selected are based on the thesis that the competing crops are important to the extent these crops occupy the area. Rice, wheat, gram, barley, sugarcane, maize occupied respectively 19, 19, 13, 9, 6 and 5 per cent of the average gross cropped area at the initial three years, *viz*., 1956-57 to 1958-59.

Both the farm (harvest) price and wholesale price series have been used. In the wholesale price series the price of gur while in the farm harvest series the price of sugarcane have been used. In the farm harvest price series the price of paddy while in wholesale price series the price of rice have been used. These constraints were imposed by the available data.

Lagged acreage and relative lagged acreage as independent variables

The one year lagged acreage is considered as an independent variable because in the short run, this has a more dominant influence than variations in price. It is only in the long run that the farmers would be able to adjust their acreage to price changes. In some cases the influence of relative lagged acreage has been found as having a dominant influence on acreage allocation under the crops.

Rainfall an independant variable

Pre-sowing period rainfall is considered as an independent variable because it has also an important influence on acreage allocation. Inspite of an increase in irrigated area, rainfall still continues to have a dominant influence on the acreage allocation.

Gross irrigated area as an independent variable

The changes in gross irrigated area has also an important influence on acreage allocation. In most of the models gross irrigated area has been considered as an independent variable.

Relative lagged yield as an independent variable

Relative lagged yield has been also included as an independent variable because the relative yield variations are also supposed to have an influence on acreage allocation. Moreover, yield variable takes care of other inputs used in agriculture.

Time as independent variable (T)

The main reason in using a trend variable rather than a more direct measure of postulated influence on supply is generally resorted to because of difficulties in obtaining reliable data for the factor in question. Sometimes several such influences are related to output and may be inter-related among themselves that their separate inclusion might involve problems of multi-collinearity in the estimation process. In the preliminary analysis the time variable T was used but its inclusion in the model was plagued by multi-collinearity problem. The varible T was excluded from the final models on this account only.

Format of the Model as used in the Study

The basic model used, is Nerlovian 'adjustment' model*

(1) $X_t^* = a + bP_{t-1} + cY_{t-1} + gZ_{t-1} + hW_t + u_t$

(2) $X_t - X_{t-1} = B\,(X_t^* - X_{t-1})$

where

X_t^* area that the farmers would plant in the year t if there were no difficulties of adjustment to changes in the price and other relevant variables.

X_t area actually planted to the crop in year t.

P_t deflated farm harvest/wholesale price of the crop in the year t.

Y_t relative yield of the crop in the year t.

Z_t Gross irrigated area in all the crops in year t.

W_t Pre-sowing period rainfall in the year t.

u_t Stochastic disturbance term

B is the coefficient of adjustment. The farmers, it is postulated, are able to increase the area of a crop in any year 't' only to the extent of a fraction B of the difference between the area they would like to plant and the area actually planted in the preceding year t-1.

Solving (1) and (2) equations the equation (3) is obtained and it is used for estimating the unknown coefficients by the method of Least Squares.

(3) **$X_t = a_0 + b_2P_{t-1} + b_3Y_{t-1} + b_4Z_{t-1} + b_5W_t + b_6X_{t-1} + V_t$

where

$a_0 = aB,\ b_2 = bB,\ b_3 = cB,$
$b_4 = gB,\ b_5 = hB,\ b_6 = 1 - B$
and $V_t = Bu_t$

In the case of certain selected crops, "the shifter" variables Y, Z, W have been omitted from the estimating equations, if they are found on preliminary analysis not important factors determining the area of that particular crop. Thus, in the case of certain

*As mentioned earlier the same model was used by Raj Krishna also.

**Details of obtaining the Reduced Form and Koyeks' model are given in the Appendix 5.1.

crops, the following estimation equation has been used

(4) $X_t + a_0 + b_2 P_{t-1} + b_6 X_{t\ 1}$

A detailed format of the estimation equations with reference to rice, wheat and sugarcane is given separately.

Symbols used in the estimating equations

X_{R_t} Area planted under rice in year '*t*'

X_{WH_t} Area sown under wheat in year *t*.

X_{S_t} Area sown under sugarcane in year *t*.

$P_{R_{t-1}} / P_{SM_{t-1}}$ Price of rice deflated by the composite index of sugarcane and maize

W_{PS} Pre-sowing period rainfall

$Z_{t\ 1}$ Gross irrigated area in year *t*-1.

$Y_{WH_{t-1}} / Y_{SBG_{t-1}}$ Yield of wheat deflated by the composite index of yield of sugarcane, barley and gram.

$P_{S_{t-1}} / P_{WHR_{t-1}}$ Price index of sugarcane deflated by the composite price index of rice and wheat.

Other variables have usual meanings with reference to the concerned crops.

The theoretical format of the models along with the estimated values of the coefficients with reference to rice, wheat and sugarcane is presented below :

Acreage response equations of Rice

(*a*) *Linear Models*

MODEL I

$$X_{R_t} = b_0 + b_2\, W_{PS} + b_3\, X_{R_{t-1}} + u$$

$$X_{R_t} = 19.27 + \underset{(3.06)}{0\,07}\ W_{PS} + \underset{(7.66)}{0.74}\ X_{R_{t-1}}$$

MODEL II

$$X_{R_t} = b_0 + b_2\, P_{R_{t-1}} / P_{SM_{t-1}} + b_3\, X_{R_{t-1}} / X_{SM_{t-1}} + u$$

$$X_{R_t} = 27.57 + \underset{(2.49)}{0.22}\ P_{R_{t-1}} / P_{SM_{t-1}} + \underset{(3{\cdot}06)}{0.52}\ X_{R_{t-1}} / X_{SM_{t-1}}$$

MODEL III

$$X_{R_t}=b_0+b_2 P_{R_{t-1}}/P_{SM_{t-1}}+b_3 W_{PS}+b_3 Z_{t-1}+u$$

$$X_{R_t}=97.17+\underset{(3.42)}{0.19} P_{R_{t-1}}/P_{SM_{t-1}}+\underset{(3.32)}{0.10} W_{PS}-\underset{(6.45)}{0.22} Z_{t-1}$$

MODEL IV

$$X_{R_t}=b_0+b_2 W_{PS}+b_3 X_{R_{t-1}}+u$$

$$=4.78+\underset{(2.43)}{\cdot 05} W_{PS}+\underset{(12.11)}{0.89} X_{R_{t-1}}$$

(*b*) *Non-Linear Models*

MODEL I

$$X_{R_t}=b_0+b_2 W_{PS}+b_3 X_{R_{t-1}}+u$$

$$X_{R_i}=0.33+\underset{(3.35)}{0.86} W_{PS}+\underset{(8.40)}{0.75} X_{R_{t-1}}$$

MODEL II

$$X_{R_t}=b_0+b_2 P_{R_{t-1}}/P_{SM_{t-1}}+b_3 W_{PS}+b_4 Z_{t-1}+u$$

$$=1.96+\underset{(2.60)}{0.18} P_{R_{t-1}}/P_{SM_{t-1}}+\underset{(2.05)}{\cdot 09} W_{PS}-\underset{(5.07)}{0.24} Z_{t-1}$$

MODEL III

$$X_{R_t}=b_0+b_2 P_{R_{t-1}}/P_{SM_{t-1}}+b_3 X_{R_{t-1}}/X_{SM_{t-1}}+u$$

$$=0.50+\underset{(2.70)}{0.22} P_{R_{t-1}}/P_{SM_{t-1}}+\underset{(3.21)}{0.54} X_{R_{i-1}}/X_{SM_{t-1}}$$

MODEL IV

$$X_{R_t}=b_0+b_2 W_{PS}+b_3 X_{R_{t-1}}+u$$

$$=0{\cdot}69+\underset{(2.79)}{0.71} W_{PS}+\underset{(13.46)}{0.89} X_{R_{t-1}}$$

MODEL V

$$X_{R_t}=b_0+b_2\, P_{R_{t-1}}/P_{SM_{t-1}}+b_3\, W_{PS}+b_4\, Z_{t-1}+u$$

$$=2.09+\underset{(1.40)}{0.10}\, P_{R_{t-1}}/P_{SM_{t-1}}+\underset{(3.00)}{0.14}\, W_{PS}-\underset{(5.60)}{0.28}\, Z_{t-1}$$

MODEL IV

$$X_{R_t}=b_0+b_2\, P_{R_{t-1}}/P_{SM_{t-1}}+b_3\, W_{PS}+b_4\, X_{R_{t-1}}/X_{SM_{t-1}}$$

$$=0.0043+\underset{(2.36)}{0.18}\, P_{R_{t-1}}/P_{SM_{t-1}}+\underset{(2.20)}{0.11}\, W_{PS}$$

$$+\underset{(4.80)}{0.70}\, X_{R_{t-1}}/X_{SM_{t-1}}$$

Acreage Response of Wheat

(*a*) *Linear Models*

MODEL I

$$X_{WH_t}=b_0+b_2\, Y_{WH_{t-1}}/Y_{SBG_{t-1}}+b_3\, X_{WH_{t-1}}+u$$

$$X_{WH_t}=-19.97+\underset{(2.24)}{0.30}\, Y_{WH_{t-1}}/Y_{SBG_{t-1}}\ \underset{(14.66)}{0.92}\, X_{WH_{t-1}}$$

MODEL II

$$X_{WH_t}=b_0+b_2\, Y_{WH_{t-1}}/Y_{SBG_{t-1}}+b_3\, X_{WH_{t-1}}/X_{SBG_{t-1}}+u$$

$$X_{WH_t}=-30.84+\underset{(1.64)}{0.44}\, Y_{WH_{t-1}}/Y_{SBG_{t-1}}$$

$$+\underset{(10.57)}{0.91}\, X_{WH_{t-1}}/X_{SBG_{t-1}}$$

MODEL III

$$X_{WH_t}=b_0+b_2\, Y_{WH_{t-1}}/Y_{SBG_{t-1}}+b_3\, Z_{t-1}+b_3\, X_{WH_{t-1}}+u$$

$$=-15.11+\underset{(1.99)}{0.23}\, Y_{WH_{t-1}}/Y_{SBG_{t-1}}+\underset{(2.58)}{0.29}\, Z_{t-1}$$

$$+\underset{(6.00)}{0.65}\, X_{WH_{t-1}}$$

MODEL IV

$$X_{WH_t} = b_0 + b_2 Z_{t-1} + b_3 X_{WH_{t-1}} / X_{SBG_{t-1}} + u$$

$$= -20.11 + \underset{(3.18)}{0.66} Z_{t-1} + \underset{(4{\cdot}08)}{0.58} X_{WH_{t-1}} / X_{SBG_{t-1}}$$

(*b*) *Non-Linear Models*

MODEL I

$$X_{WH_t} = b_0 + b_2 Y_{WH_{t-1}} / Y_{SBG_{t-1}} + b_3 X_{WH_{t-1}} + u$$

$$X_{WH_t} = -0{\cdot}34 + \underset{(2.26)}{0{\cdot}26} Y_{WH_{t-1}} / Y_{SBG_{t-1}} + \underset{(13.61)}{0{\cdot}91} X_{WH_{t-1}}$$

MODEL II

$$X_{WH_t} = b_0 + b_3 Y_{WH_{t-1}} / Y_{SBG_{t-1}} + b_3 X_{WH_{t-1}} / X_{SBG_{t-1}} + u$$

$$= -0.48 + \underset{(1.73)}{0.35} Y_{WH_{t-1}} / Y_{SBG_{t-1}} \underset{(10.73)}{0.90} X_{WH_{t-1}} / X_{SBG_{t-1}}$$

MODEL III

$$X_{WH_t} = b_0 + b_2 Y_{WH_{t-1}} / Y_{SBG_{t-1}} + b_3 Z_{t-1} + b_4 X_{WH_{t-1}} + u$$

$$X_{WH_t} = -0{\cdot}26 + \underset{(2{\cdot}21)}{0.22} Y_{WH_{t-1}} / Y_{SBG_{t-1}} + \underset{(3{\cdot}23)}{0\ 32} Z_{t-1} + \underset{(6.31)}{0.61} X_{WH_{t-1}}$$

MODEL IV

$$X_{WH_t} = b_0 + b_2 Y_{WH_{t-1}} / Y_{SBG_{t-1}} + b_3 - Z_{t-1} + b_4 X_{WH_{t-1}} / X_{SBG_{t-1}} + u$$

$$= -0{\cdot}67 + \underset{(1.82)}{0.24} Y_{WH_{t-1}} / Y_{SBG_{t-1}} + \underset{(3{\cdot}09)}{0.50} Z_{t-1} + \underset{(4.50)}{0.55} X_{WH_{t-1}} / X_{SBG_{t-1}}$$

Acreage Response of Sugarcane

(*a*) *Linear Models*

MODEL I

$$X_{S_t} = b_0 + b_2 P_{S_{t-1}} / P_{WHR_{t-1}} + b_3 X_{S_{t-1}} + u$$

$$X_{S_t} = 3.21 + \underset{(3.64)}{0.26} P_{S_{t-1}} / P_{WHR_{t-1}} + \underset{(4.19)}{0.71} X_{S_{t-1}}$$

MODEL II

$$X_{S_t} = b_0 + b_2 P_{S_{t-1}} / P_{WHR_{t-1}} + b_3 X_{S_{t-1}} / X_{WHR_{t-1}} + 4$$

$$X_{S_t} = -17.60 + \underset{(2.66)}{0\ 24} P_{S_{t-1}} / P_{WHR_{t\ 1}} + \underset{(4.32)}{0.89} X_{S_{t-1}} / X_{WHR_{t-1}}$$

MODEL III

$$X_{S_t} = b_0 + b_2 P_{S_{r-1}} / P_{WHR_{t-1}} + b_3 X_{S_{t-1}} + 4$$

$$X_{S_t} = -6.69 + \underset{(3.68)}{0.23} P_{S_{t-1}} / P_{WHR_{t-1}} + \underset{(6\ 00)}{0.81} X_{S_{t-1}}$$

MODEL IV

$$X_{S_t} = b_0 + b_2 P_{S_{t-1}} / P_{WHR_{t-1}} + b_3 X_{S_{t-1}} / X_{WHR_{t-1}} + u$$

$$= -1.72 + \underset{(2.37)}{0.19} P_{S_{t-1}} / P_{WHR_{t-1}} + \underset{(3.68)}{0.76} X_{S_{t-1}} / X_{WHR_{t-1}}$$

(*b*) *Non-Linear Models*

MODEL I

$$X_{S_t} = b_0 + b_2 P_{S_{t-1}} / P_{WHR_{t-1}} + b_3 X_{S_{t-1}} + u$$

$$X_{S_t} = 0{\cdot}07 + \underset{(4.47)}{0.30} P_{S_{t-1}} / P_{WHR_{t-1}} + \underset{(4.45)}{0.67} X_{S_{t-1}}$$

MODEL II

$$X_{S_t} = b_0 + b_2 P_{S_{t-1}} / P_{WHR_{t-1}} + b_3 X_{S_{t-1}} / X_{WHR_{t-1}} + u$$

$$X_{S_t} = -0.59 + \underset{(3.86)}{0.37} P_{S_{t-1}} / P_{WHR_{t-1}} + \underset{(5.41)}{0.92} X_{S_{t-1}} / X_{WHR_{t-1}}$$

MODEL III

$$X_{S_t} = b_0 + b_2\, P_{S_{t-1}} / P_{WHR_{t-1}} + b_3\, X_{S_{t-1}} + u$$

$$X_{S_t} = -0{\cdot}24 + \underset{(3{\cdot}91)}{0{\cdot}32}\, P_{S_{t-1}} / P_{WHR_{t-1}} + \underset{(6{\cdot}0)}{0.80}\, X_{S_{t-1}}$$

MODEL IV

$$X_{S_t} = b_0 + b_2\, P_{S_{t-2}} / P_{WHR_{t-2}} + b_3\, X_{S_{t-1}} + 4$$

$$= 0.43 + \underset{(2.81)}{0.26}\, P_{S_{t-2}} / P_{WHR_{t-1}} + \underset{(3.35)}{0.51}\, X_{S_{t-1}}$$

MODEL V

$$X_{S_{t-1}} = b_0 + b_2\, P_{S_{t-1}} / P_{WHR_{t-1}} + b_3\, X_{S_{t-1}} / X_{WHR_{t-1}} + 4$$

$$X_{S_t} = -0{\cdot}29 + \underset{(3.0)}{0.32}\, P_{S_{t-1}} / P_{WHR_{t-1}} + \underset{(4.1)}{0.30}\, X_{S_{t-1}} / X_{WHR_{t-1}}$$

The response coefficients of rice, wheat and sugarcane obtained on the basis of above estimating equations are presented in a consolidated form in tables 5.1, 5.2, 5.3 respectively.

Comments on the results

Rice

The lagged and relative lagged acreage both have influenced allocation of area under rice. In fact, the influence of these two variables has been very dominant. Pre-sowing rainfall also affected the area allocation under rice. Out of ten models considered, the influence of price on acreage allocation under rice is corroborated by the results obtained from six models only. It is surprising to note that an increase in gross irrigated area has shown a negative influence on the area allocation under rice. The effect of yield on acreage allocation under rice did not turn up to be significant in any of the models considered above.

Wheat

It is surprising to note that the effect of price variable did not turn out to be significant in any of the models considered for estimation of supply response. Pre-sowing rainfall also did not seem to have

TABLE 5.1. **Acreage response**

	Models	Constant	Relative Price	Relative Yield
	(1)	(2)	(3)	(4)
Linear				
Using FHP	Model I	19.27	...	...
	Model II	27.57	0.2230 (2.4852)*	...
Using Whole-sale Prices	Model III	97.17	0.1880 (3.4213)*	...
	Model IV	4.78	...	...
Non-Linear				
Using FHP	Model I	0.3332	...	...
	Model II	1.9598	0.1752 (2.6015)*	...
	Model III	0.4967	.2215 (2.6964)*	...
Using Whole-sale Prices	Model IV	0.6941	...	...
	Model V	2.0936	0.09733 (1.4028)***	...
	Model VI	0.0043	0 1817 (2.3622)*	...

* Significant at 5% level.
** Significant at 10% level.
*** Significant at 20% level.

equations of Rice

Gross Irrigated Area	Pre-Sowing Rainfall	Lagged Acreage	Relative Lagged Acreage	R^2	$\bar{R}^2$	Durbin Watson Statistics
(5)	(6)	(7)	(8)	(9)	(10)	(11)
...	0.0708 (3.0607)*	0.7422 (7.6586)*	...	0.81	0.79	2.555
...	...	...	0.5248 (3.0551)*	0 49	0.43	...
−0.2200 (6.4530)*	0.0991 (3.3189)*	—	...	0.78	0.74	1.5879
...	0 0539 (2.4267)*	0 8928 (12.1164)*	...	0.88	0.87	2.5386
...	0.8637 (3.349)*	0.7479 (8.4041)*	...	0.84	0.82	2.5017
−0.2409 (5.0727)*	0 0940 (2.0460)**	—	...	0.71	0.65	...
...	...	...	0.5357 (3.2087)*	0.52	0.46	—
—	0.7104 (2.7859)*	0.8927 (13.4639)*	...	0.90	0.89	2.4487
−0.2825 (5.6010)*	0.1427 (3.0047)*	...	...	0.67	0.62	1.6163
...	0.1134 (2.1997)*	—	0.6971 (4.8001)*	0.61	0.55	2.1814

TABLE 5.2. **Acreage response**

	Models	Constant	Relative Price	Relative Yield
	(1)	(2)	(3)	(4)
Linear				
Using FHP	Model I	−19.97	...	0.2982 (2 2438)*
	Model II	−30.84	...	0 4369 (16.372)***
Using WSP	Model III	−15.10	...	0 2324 (1.9854)**
	Model IV	−20.11	...	...
Non-Linear				
Using FHP	Model I	−0.3374	...	0.2635 (2.2595)*
	Model II	−0.4804	...	0.3464 (1.7255)**
Using WSP	Model III	−0.2628	...	0.2158 (2.2104)*
	Model IV	−0.6701	...	0.2350 (1.8152)**

* Significant at 5% level.
** ,, 10% ,,
*** ,, 20% ,,

equations of wheat

Gross Irrigated Area	Pre-Sowing Rainfall	Lagged Acreage	Relative Lagged Acreage	R^2	$\bar{R}^2$	Durbin Watson Statistics
(5)	(6)	(7)	(8)	(9)	(10)	(11)
...	...	0 9240 (14.6595)*	...	0.96	0.96	2.0680
...	...	...	0.9096 (10.5710)*	0.94	0.93	2.294
0.2948 (2.5771)*	...	0.6527 (6.0023)*	...	0.97	0.97	2.410
0.6561 (3.1770)*	...	...	0.5770 (4.0836)*	0.96	0.95	2.116
...	...	0 9096 (13.6101)*	...	0.96	0.96	1.987
...	...	...	0.9004 (10.7255)*	0.94	0.93	...
0.3155 (3.226)*	...	0.6061 (6.3141)*	...	0.97	0.97	2.497
0.5041 (3.089)*	...	...	0.5523 (4.5048)*	0.96	0.96	2.617

TABLE 5.3. **Acreage response**

	Models	Constant	Relative Price	Relative Yield
	(1)	(2)	(3)	(4)
Linear				
Using FHP	Model I	3.2132	0.2555 (3.6428)	...
	Model II	−17.5991	0.2416 (2.6567)*	...
Using WSP	Model III	−6.6881	.2254 (3.679)*	...
	Model IV	−1.7199	0 1855 (2.3717)*	...
Non-Linear				
Using FHP	Model I	0.0662	0.3001 (4.4682)	...
	Model II	−0.5867	0 3653 (3.8613)*	...
Using WSP	Model III	−0.2407	0.3151 (3.9081)*	...
	Model IV	0.4315	0.2645 (2.8073)*	...
	Model V	−0.2892	0,3238 (3.0085)*	...

Note . In Non-Linear Model IV two years' time lag has been used.
*** Significant at 5% level.**

equations of sugarcane

Gross Irrigated Area	Pre-Sowing Rainfall	Lagged Acreage	Relative Lagged Acreage	R^2	$\bar{R}^2$	Durbin Watson Statistics
(5)	(6)	(7)	(8)	(9)	(10)	(11)
...	...	0.7084 (4.1899)*	...	0.57	0.52	8.8110
...	...	...	0.8918 (4.3232)*	0.52	0.47	1.8430
...	...	...	0.8104 (6.0003)*	0.67	0.64	1.4903
...	...	...	0.7606 (3.6824)*	0.41	0.35	1.4277
...	...	0.6667 (4.4482)*	...	0 64	0·60	1.8642
...	...	...	0.9185 (5.4163)*	0.64	0.60	1.9785
...	...	0.7975 (6.0361)*	...	0.68	0.65	1.4614
...	...	0.5131 (3.3470)*	...	0.59	0.55	1.4275
...	...	...	0 3029 (4.1006)*	0.47	0.41	1.4446

TABLE 5.4. **Estimated Coefficient of Adjustment**

S. No.	Crop.	Linear Models				Non-Linear Models					
		I	II	III	IV	I	II	III	IV	V	VI
1.	Rice	0.2578	0.4752	..	0 1072	0.2521	..	.4643	.1073	..	.3029
2.	Wheat	0.0760	0.0904	0·3473	.4230	.0904	0.0996	.3979	..	..	, .
3.	Sugarcane	0.2916	0.1082	0,1896	.2394	0.3333	.0815	.2025	.4831	.6971	..

Source : Tables 5.1, 5.2, 5.3.

any influence on the allocation of acreage under the crop. Yield variable exhibited a dominant influence on acreage allocation under the crop. The break-through in agriculture brought about by Green Revolution led to increase in yield of the crop. It can be safely inferred that the increased productivity per hectare has positively influenced the acreage allocation under the crop. The effect of gross irrigated area has also influenced the area under wheat cultivation.

Sugarcane

The influence of relative yield, gross irrigated area and pre-sowing rainfall did not turn out to be significant on acreage allocation of sugarcane. Lagged acreage and price demonstrated a positive influence on the allocation of land under sugarcane cultivation. The price response of the farmers with reference to allocation of land for cultivation of sugarcane has been emphatically demonstrated by the results presented in table 5.3. The impact of price lagged by two years was also seen on the acreage allocation under the crop. It may be most probably due to ratooning followed in sugarcane cultivation. The coefficients of adjustment* are presented in a consolidated form in table 5.4.

The low values of adjustment coefficients clearly indicate that the influence of price and other shifter variables on acreage allocation under the crop. *viz.*, rice, wheat and sugarcane was not felt in close proximity. The adjustment coefficients (B's) presented above have been worked out on the basis of following relationship.

$B = 1 -$ value of coefficient of lagged or relative lagged acreage.

* The values of coefficients of adjustment for equation 2 of Nerlovian Model have been obtained from the estimating equations presented earlier.

REFERENCES

1. Dorfman, Robert, Prices and markets, Otto Eckstein (ed.) *Foundation of Modern Economic Series*, Prentice Hall, Inc. Engle Wood Cliffs, New Jersey, 1967, p. 136.
2. Found, William C., *A Theoretical Approach to Rural Land Use Patterns*. Edward Arnold (Publishers) Ltd., London, 1971, p. 22.

3. J.K. Galbraith and J.D. Black, The maintenance of agricultural production during depression : explanations reviewed, *Journal of Political Economy*, June 1938, p. 307.
4. B.D. Giles, Agriculture and price mechanism, in *Oxford Studies in the Price Mechanism*, edited by T. Wilson and P.W.S. Andrews (Oxford, 1951), pp. 173-4.
5. Marc Nerlove, The dynamics of supply : Estimation of farmers' response to price, The John Hopkins Press, 1958.
6. Indian Council of Social Sciences Research : A Survey of Research in Economics, Vol. 3 Agriculture, Part I, Chapter IV, Allied Publishers, Bombay, 1975.
7. Dharm Narain, *Impact of Price Movement on Areas Under Selected Crops in India, 1900-1939*, Cambridge University Press, 1965.
8. Raj Krishna, Farm supply response in India Pakistan : A case Study of the Punjab region. *The Economic Journal*, Vol. 73, No. 291, September 1963, pp. 477-87.
9. G. Parthasarthy, Substitution between sugarcane and paddy in Madras state, *Indian Journal of Agricultural Economics*, Vol. 14, No. 3, July-September, 1959, pp. 31-39.
10. P.V. John, Responsiveness of prices in Uttar Pradesh, 1954-63, *Indian Journal of Agricultural Economics*, Vol. 20, No. 1, January-March, 1966, pp. 40-47.
11. S.C. Gupta and A. Majid, '*Farmers' Response to Prices and Marketing Policies affecting Sugarcane and Paddy : A Case Study of Eastern U.P.*, 1965.
12. Chandresh Kumar "Farmers' response to price changes of sugarcane in Uttar Pradesh, 1951-52 to 1966-67, *Economic and Political Weekly*, Vol. 5, No. 18, May 2, 1970.
13. Y. Satyanarayan, Factors affecting sugarcane acreage in India, *Indian Journal of Agricultural Economics*, Vol. 22, No. 2, April-June, 1967.
14 J.R. Behrman, *Supply Response in Underdeveloped Agriculture*, North Holland, 1968.
15. Hossein Askari and John Thomas Cummings, Estimating agricultural supply response with Nerlove Model : A survey, *International Economic Review*, Vol. 18, No. 2, June 1977.
16. Marc Nerlove ; *The Dynamics of Supply : Estimation of Farmers' Response to Price*, The John Hopkins Press, 1958.

6

Projections of Food Availability

In the world there are only six countries viz., the Union of Soviet Socialist Republic, the United States of America, China, Japan, Brazil and Indonesia which have more population than the State of Uttar Pradesh. The State has a challenging problem of feeding the rapidly growing population. There is also a need to increase the food intake of most of the population both in quantity and quality terms. The food intake of at least some of the population is so restricted that the calorie intake is insufficient for full growth and activity. The diet of a large number of people is lacking in important qualitative features such as amino acids and vitamins. As a result, health is impaired, resistance to disease is lowered and capacity to work is reduced. At this low level of food intake the situation gets worsened still further when more numbers are added to the existing population without any corresponding increase in food production.

According to population census of 1971, the State occupies the first place in the country and more than 16 per cent of the population of the country resides in the State. During the two decades 1951-71 the percentage share of the State in the population of the country decreased from 17.5 per cent to 16.1 per cent. The decennial growth rates, size and density of population of the State and the country are presented in Table 6.1.

It can be seen from Table 6.1 that during 1951-71 the decennial growth rates of population in the State were lower than the corresponding growth rates in the country whereas the density of population of the State remained at a higher level throughout this time period, compared to the corresponding

TABLE 6·1 : **Growth of population in Uttar Pradesh and India : 1921-71**

		U.P.			India		
S.No·	Year	Population (lakhs)	Decennial Growth rate (Per cent)	Density of population	Population	Decennial Growth rate (Per cent)	Density of population
(1)	(2)	(3)	(4)	(5)	(6)	(7)	(8)
1.	1921	467	..	159	2514	..	77
2.	1931	498	6.7	169	2790	11.0	85
3.	1941	565	13.6	192	3187	14.2	98
4.	1951	632	11.8	215	3611	13.3	111
5.	1961	737	16.7	251	4391	21.6	134
6.	1971	883	19.8	300	5479	24.8	178

Source : Economic Survey of Uttar Pradesh, 1977-78.

density in the country. According to Census 1971, the density of population of the State was 68.5 per cent higher than the corresponding density of the country. During the two decades 1951-71 the density of population in the State increased by 39.5 per cent whereas the corresponding increase in the density of population of the country was 60.4 per cent. The population of the State and the country as a whole increased by 39.7 per cent and 51.7 per cent respectively during the two decades 1951-71. From the population statistics presented in the above table, it is evident that the pressure of population has been mounting in every decade.

The table 6.2 presents the quinquennial total and per capita production of cereals, pulses and foodgrains in the State during 1950-51 to 1975-76.

During 1950-51 to 1970-71 the production of total foodgrains in the State increased by 66.3 per cent while the corresponding increase in the production of cereals and pulses was 88.7 per cent and 1.5 per cent respectively. During the quinquennium 1970-71 to 1975-76 the total foodgrain production decreased by 0.6 per cent whereas the corresponding decrease in the production of

TABLE 6.2 : **Production of food-grains in Uttar Pradesh, 1950-51 to 1975-76**

S.No.	Year	Total production (000 tonnes)			Per Capita Production (Kilogrammes)		
		Cereals	Pulses	Food Grains	Cereals	Pulses	Food Grains
1.	1950 51	8751	3024	11775	138	48	186
2.	1955-56	9138	2920	12058	136	43	179
3.	1960-61	10662	3824	14486	145	52	197
4.	1965-66	10014	3277	13291	124	41	165
5.	1970-71	16516	3069	19585	187	35	222
6.	1975-76	16820	2657	19477	174	28	202
	Average	11983	3129	15112	151	41	192

Source : Estimates of Area, Production and Yield of principal Crops in India ; Directorate of Economics and Statistics, Ministry of Agriculture and Irrigation, Government of India, New Delhi.

pulses was 13.4 per cent. During 1950-51 to 1970-71 the per capita production increased by 19.4 per cent whereas the corresponding increase in per capita cereals was 35.5 per cent. The per capita production of pulses decreased by 27.1 per cent during this period. In 1975-76 the per capita production of cereals, pulses and food-grains decreased by 7.5, 25.0 and 9.0 per cent respectively, compared to 1970-71. It is quite evident that the production of pulses declined considerably throughout the period of investigation.

ESTIMATION OF FUTURE FOODGRAIN AVAILABILITY

The principal objective of planned economic development of the State is raising the standard of living of its people and it can not be achieved without providing for their feeding at an adequate level. The need for careful assessment of future total foodgrain availability in the State on the basis of present per capita foodgrain availability need not be over-emphasized. Such projections of total foodgrain availability have a limited purpose of ascertaining roughly the total foodgrain availability in the State in the next decade on the assumption that the present level of per capita availability of foodgrains is maintained. It is admitted without any reservation that the sophisticated nutritional approach or demand analysis is beyond the scope of the present investigation.

PER CAPITA NET AVAILABILITY OF FOODGRAINS

Net availability stands for the quantity of foodgrains available for human consumption. The per capita net availability of cereals, pulses and foodgrains has been estimated from the Statistics of production presented in Table 6.2. While estimating the net availability it is assumed that the whole production is consumed within the State and nothing is exported or imported. Also, an allowance of 12.5 per cent has been made for seed, feed requirements and wastage. Net availability of foodgrains is presented in Table 6·3. The per capita net availability of cereals,

TABLE 6.3 : **Net availability of food-grains in Uttar Pradesh, 1950-51 to 1975-76**

		Net Availability (000 tonnes)			Per Capita Net Availability (Kilogrammes)		
S.No.	Year	Cereals	Pulses	Food Grains	Cereals	Pulses	Food Grains
1.	1950-51	7657	2646	10303	121	42	163
2.	1955-56	7996	2555	10551	119	38	157
3.	1960-61	9329	3346	12675	127	45	172
4.	1965-66	8762	2867	11630	108	36	144
5.	1970-71	14452	2685	17137	164	30	194
6.	1975-76	14718	2325	17042	152	24	176
	Average	10486	2737	13223	132	36	168

pulses and total foodgrains in 1950-51 was 121, 42, 163 kilogrammes per annum respectively. There was an increase of 25.6 and 8.0 per cent in the net per capita availability of cereals and total foodgrains whereas the per capita net availability of pulses decreased by 42.9 per cent during the entire period of investigation. Among the years presented in the table 6.2 the highest level of per capita net availability of cereals and food-grains was achieved in 1970-71 whereas the highest level of per capita of net availability of pulses was recorded in 1960-61. In 1975-76 the per capita net availability of cereals, pulses, food-grains declined and the decline was considerable in case of pulses compared to 1970-71.

RATIO OF NET AVAILABILITY OF UTTAR PRADESH TO INDIA

The ratio of per capita net availability of cereals, pulses and total foodgrains is presented in Table 6.4.

TABLE 6.4 : **Ratio of per capita net availability of cereals, pulses and total food-grains in Uttar Pradesh to India, 1950-51 to 1975-76**

S.No.	Year	Cereals	Pulses	Total foodgrains
1.	1950-51	0.99	1.91	1.13
2.	1955-56	0.90	1.46	0.99
3.	1960-61	0.87	1 80	1.01
4.	1965-66	0.82	2.00	0.97
5.	1970-71	1 08	1.58	1.13
6.	1975 76	1.03	1.26	1 06
7.	Average	0.96	1 64	1.05

Inspite of the fact that the per capita net availability of pulses decreased considerably during the whole period under investigation. The per capita net availability of pulses remained at a higher level throughout the period in the State compared to All India. During the last fifteen years the position of State compared to All India with reference to net availability of cereals and pulses was quite encouraging. The position of the State in per capita availability of cereals compared to All India remained below in 1950-51, 1955-56, 1960-61 and 1965-66. An emerging conclusion is that the per capita net availability of total food-grains remained either at a very close level or much higher level than All India.

Projections of cereals, pulses and total food-grains

The projections of cereals, pulses and total food-grains on the basis of average* per capita net availability are presented in Table 6.5.

TABLE 6.5 : **Projected food availability, 1980-81 to 1990-91**

S.No.	Commodities	Average Net Availability (Kilogrammes)	Projected Total Food Availability ('000 tonnes) 1980-81	1985-86	1990-91
1.	Cereals	132	13956	15146	16119
2.	Pulses	36	3806	4131	4396
3.	Total Food Grains	168	17762	19277	20515

On the basis of the assumption that the average net availability would be maintained in the next decade also the total availability of cereals would be 13956, 15146, 16119 thousand tonnes in 1980-81, 1985-86 and 1990-91 respectively and the corresponding availability of pulses would be 3806, 4131 and 4396 thousand tonnes respectively. The net availability of total food-grains has been obtained by adding the net availability of cereals and pulses.

PROJECTIONS OF NET AVAILABILITY OF CEREALS, PULSES AND TOTAL FOOD-GRAINS BASED ON THE HIGHEST (H) AND THE LOWEST (L) PER CAPITA NET AVAILABILITY

It is seen from Table 6.6 that the highest per capita net availability of cereals, pulses and total food-grains was 164, 45

* Average per capita net availability has been computed on the basis of per capita net availability for the years presented in the table 6.3.

and 194 kilogrammes respectively whereas the lowest per capita net availability of cereals, pulses and total food-grains was 108, 24 and 144 kilogrammes respectively. On the basis of the highest and the lowest per capita net availability the total net availability of cereals, pulses and food-grains has been estimated.

TABLE 6.6 : **Projected net availability of cereals, pulses and food-grains, 1980-81 to 1990-91**

S.No.	Commodities	Net avail-ability (kgs.)		1980-81		1985-86		1990-91	
		H	L	H	L	H	L	H	L
1.	Cereals	164	108	17340	11419	18818	12392	20027	18188
2.	Pulses	45	24	4758	2537	5163	2754	5495	2931
3.	Total Food Grains	194	144	20511	15225	22260	16523	23690	17584

It can be seen from the Table 6.6 that the highest estimated net availability of cereals, pulses and total foodgrains would be 20511, 22260, 23690 thousand tonnes in the years 1980-81, 1985-86 and 1990-91 respectively. Basing the projections on the lowest corresponding net availability the estimated net availability would be 15225, 16523, 17584 thousand tonnes. All the projections are based on the assumption that while maintaining the present net availability the projected net availability of cereals, pulses and total foodgrains would be of the magnitudes presented in Tables 6.5 and 6.6. It may be noted that the net availability of cereals and pulses would not be equal to net availability of total food-grains simply because of the fact that the net availability of cereals, pulses and total food-grains are not based on the net availability of the same year.

7

Summary[*] and Conclusions

AGRICULTURAL development plays an important role in the economic development of the country. The economists, sociologists, political thinkers, planners and administrators hold the view that the success of economic development programmes depends ultimately on the agricultural development. India is a vast country comprising of 22 States and 9 Union Territories. The level of economic development achieved in these various political cum-administrative subdivisions finally determines the economic development of the country. Keeping in view the magnitude of population and vastness of area the economic development of Uttar Pradesh has an important role to play in the economic development of the country. In the economic development of the State the role of agricultural development is quite decisive.

During the post-independence period the quality of available statistical data has certainly improved due to use of latest statistical techniques of data collection. While the quality of statistical data has improved gradually, the comparability of data over time has been lost. The accuracy of results obtained from the investigation certainly depends on the quality of data used. The research worker has to content himself by assuming that reasonable level of accuracy was maintained in collection of data by the agencies responsible for its collection.

Broad trends in area, production and yield of the principal crops grown in the State have a direct bearing on the prospective agricultural development of the State.

---

* Detailed summary is enclosed separately.

The study of broad trends of area, production and yield of rice, wheat, sugarcane, barley, gram and maize revealed that there was an increase both in area and production of rice, wheat, sugarcane and maize but both the area and production of barley and gram declined during the period of investigation. The yield of all the selected crops maintained an overall positive rate of growth.

Cropping pattern in the State has undergone a change. The percentage share of rice in gross cropped area remained around 19 per cent, while the increase in percentage share of wheat in gross cropped area was considerable. The crop of wheat improved its position in percentage share of gross cropped area from 17.3 per cent to 26.3 per cent. There was a marked decrease in the area under pulses and the share of pulses in gross cropped area declined from 21.5 to 15.1 per cent. An emerging conclusion is that there is a need for immediate attention to increase production of pulses either through expansion of area or through intensive cultivation. Increased demand for pulses is to be met through increased production only.

A review of land use pattern in the State revealed that the area under current fallows and other fallows declined. It is an indication of horizontal expansion of area under cultivation. The per capita area sown has declined and the chances for horizontal expansion of land are limited. Therefore, the increase in production can be obtained only through vertical expansion of land. Increase in area under assured irrigation, use of chemical fertilizers and high yielding varieties in the State is a step in this direction. The disparities present in month-wise and year-wise distribution revealed erratic occurrence of rainfall. The nature of rainfall is responsible for many drought years in the State. Inspite of increase in area under assured irrigation, the importance of rainfall in agriculture has not diminished.

The general price level is mostly determined by the behaviour of agricultural prices. Therefore, the price fluctuations in farm commodities need to be kept to minimum. The price level of farm commodities is to be controlled in such a manner that the producers get economically remunerative price for their produce and the consumers are saved from exploitation by the traders. The degree of variability in farm (harvest) price of gram was the highest while the lowest price variability was observed in the price of rice. Both the series of Farm Harvest price and Wholesale price exhibited an erratic pattern but the overall position revealed an increasing trend.

For a long time a controversy prevailed on applicability of Nerlovian Lag Adjustment Model to underdeveloped countries. A number of research studies pertaining to our country revealed the relevance of applicability of the above model to underdeveloped country like India. The present investigation revealed that the acreage response in case of rice and sugarcane with reference to price was positive but in case of wheat the effect of shifter variable like yield was more important than the price variable.

An estimate of likely total net availability of foodgrains was obtained on the assumption that the highest net availability of food-grains obtained in the past years is maintained. The projected total net availability of foodgrains for years 1980-81, 1985-86, 1990-91 worked out to 20511 thousand tonnes, 22260 thousand tonnes and 23690 thousand tonnes respectively.

There are two important policy decisions to be taken with respect to farm production in the State :

(*i*) The emphasis is to be laid on vertical expansion of land. In this connection increase in cropping intensity through the use of more chemical fertilizers and assured irrigation need special mention.

(*ii*) There is need for development of such a farm technology which keeps farms permanently under crops.

Appendices

APPENDIX TABLE 2.1

Area under selected agricultural crops in Uttar Pradesh, 1950-51 to 1977-78

('000 hectares)

Year/Crop	Rice	Wheat	Sugar-cane	Barley	Gram	Maize
(1)	(2)	(3)	(4)	(5)	(6)	(7)
1. 1950·51	3852.2	3315.7	1013 6	1947.4	2439.3	833.9
2. 1951-52	3491.3	3359.5	1199.9	1886.4	2321.2	824.0
3. 1952·53	3580.7	3375.6	1070 4	1942.9	2681.5	860.3
4. 1953-54	3663.8	3735.0	798.5	2057.4	2667.7	1021 8
5. 1954-55	3678.9	3761.7	927.4	1945.6	2657.7	1049.6
6. 1955·56	3762.6	4032 6	1100.6	1969.3	2717.2	959.7
7. 1956-57	3856.9	4038.7	1240.8	1996.7	2487.5	1014.2
8. 1957-58	3896.5	3147.3	1226 7	1776 0	2573.5	1083 2
9. 1958-59	4130.3	3870.5	1111.6	1858.1	2722 0	1093.3
10. 1959-60	4148.3	3876.4	1191.3	1845.8	2693.9	1091 7
11. 1960-61	4184.5	3938.8	1328,9	1846.3	2552 5	1054.1
12. 1961·62	4174.0	4100.0	1362.6	1824.0	2576.0	1086 0

(*Contd.*)

(1)	(2)	(3)	(4)	(5)	(6)	(7)
13. 1962-63	4278.0	4039.0	1262.0	1677.0	2468.0	1075.0
14. 1963-64	4349.0	3917.0	1229.0	1582.7	2600.0	1086.4
15. 1964-65	4449.3	3965.4	1389.0	1477.0	2627.1	1120.5
16. 1965-66	4206.8	4114.8	1489.7	1481.8	2583.7	1170.0
17. 1966-67	4445.8	4394.3	1189.7	1510.1	2383.5	1261.8
18. 1967-68	4398.3	4969.7	997.0	1612.3	2263.3	1456.0
19. 1968-69	4521.1	5239.1	1202.9	1411.2	2244.9	1502.1
20. 1969-70	4533.9	5378.4	1376.6	1484.9	2261.5	1505.3
21. 1970-71	4562.5	5907.2	1345.2	1323.1	2077.9	1508.2
22. 1971-72	4722.2	6045.9	1274.1	1312.1	1988.6	1479.0
23. 1972-73	4374.4	6135.3	1308.0	1288.0	1920.5	1482.7
24. 1973-74	4470.9	6009.1	1472.7	1268.1	1955.9	1483.2
25. 1974-75	4529.8	6152.0	1491.5	1232.0	1721.1	1394.3
26. 1975-76	4622.1	6302.1	1440.7	1202.3	1725.5	1395.5
27. 1976-77	4652.6	6624.0	1455.9	1075.2	1658.7	1322.9
28. 1977-78*	4828.0	6654.0	1691.4	975.2	1632.1	1251.1

*Provisional.

Source : Estimates of Area, Production and Yield of Principal Crops in India, Directorate of Economics and Statistics, Ministry of Agriculture and Irrigation, Government of India, New Delhi.

APPENDIX TABLE 2.2

Production of selected agricultural crops in Uttar Pradesh, 1950-51 to 1977-78

('000 tonnes)

S. No.	Year	Rice	Wheat	Gur	Barley	Gram	Maize
	(1)	(2)	(3)	(4)	(5)	(6)	(7)
1.	1950-51	1998.5	2721.1	2949.8	1712.8	1453.7	651.1
2.	1951-52	1549.6	2573.7	3220.1	1508.8	1294.0	693.0
3.	1952-53	1881.9	2869.4	2693.2	1793.6	1597.1	818 0
4.	1953-54	2291.6	3155.6	2106.5	1786.0	1768.7	673.2
5.	1954.55	2190.0	3336.7	2873.1	1741.3	1803.2	813 8
6.	1955-56	2587.9	3090.1	2987.1	1627.3	1465.8	626.5
7.	1956-57	2322.4	3164.9	3553.4	1622.0	1531.2	1000.2
8.	1957-58	2337.9	2749.5	3119.5	1280.1	1440.1	797.8
9.	1958-59	3032.3	3085.0	3125.8	1417.8	1785.6	621.1
10.	1959-60	2464.4	3293.7	3274.9	1485.7	1510.3	1029.6
11.	1960-61	3150.9	3944.6	5451.6	1687.2	1831.3	625.2
12.	1961-62	3345.0	4155.0	5126.6	1755.0	1518.0	694.0

(Contd.)

(1)	(2)	(3)	(4)	(5)	(6)	(7)
13. 1962-63	3134.0	3211.0	4320.0	1333.0	1547.0	998 0
14. 1963-64	3278.0	2715.0	4755.0	1107.9	1423 8	777.9
15. 1964-65	3323.4	4117.9	5624.0	1432.0	1926 9	888.2
16. 1965-66	2342.1	3754.7	5665.0	1342.0	1565.8	1121 0
17. 1966-67	2013.1	4230.3	3937.0	1278.5	1967.0	1076.1
18. 1967-68	3262.1	5840.7	3790.8	1766 3	1650.9	1134.4
19. 1968-69	2922.1	6086.8	5054.4	1149 5	1544.6	1338.2
20. 1969-70	3532.9	6314.3	6067.9	1467.0	1775.9	1192.2
21. 1970-71	3700.9	7689.5	5467.2	1430.5	1544.0	1796.3
22. 1971-72	3776.5	7550.1	4935.4	1359.5	1566.5	842 6
23. 1972-73	3273.0	7515.2	5672.7	1299.8	1461.2	1356.6
24. 1973-74	3858.8	5878.8	6877.3	1166.1	1044.2	1088.2
25. 1974-75	3523.4	7175.6	6147.9	1366.4	1009.6	827.3
26. 1975-76	4294.0	8552.5	3835.9	1373.7	1250.1	1014.0
27. 1976-77	4290.9	8939.9	6521.6	1167.7	1361.9	1097.0
28. 1977-78*	5141.8	9511 0	8075.6	1097.7	1197.9	940 3

*Provisional.

Source : Estimates of Area. Production and Yield of Principal Crops in India, Directorate of Economics and Statistics, Ministry of Agriculture and Irrigation, Government of India, New Delhi.

APPENDIX TABLE 2.3

Per hectare yield of selected agricultural crops in Uttar Pradesh, 1950-51 to 1977-78

(*Kilogrammes*)

S. No.	Year	Rice	Wheat	Gur	Barley	Gram	Maize
	(1)	(2)	(3)	(4)	(5)	(6)	(7)
1.	1950-51	519	821	2910	880	596	781
2.	1951-52	444	766	2684	800	557	841
3.	1952-53	526	850	2516	923	596	951
4.	1953-54	625	845	2635	868	663	659
5.	1954-55	595	887	3098	895	679	775
6.	1955-56	688	766	2714	826	539	652
7.	1956-57	602	784	2864	812	616	988
8.	1957-58	600	734	2548	721	560	737
9.	1958-59	734	797	2812	763	656	568
10.	1959-60	594	850	2749	805	561	943
11.	1960-61	753	1001	4102	914	717	593
12.	1961-62	802	1013	3764	962	589	639

(*Contd.*)

(1)	(2)	(3)	(4)	(5)	(6)	(7)
13. 1962-63	733	795	3423	795	627	928
14. 1963-64	754	693	3870	700	548	716
15. 1964-65	747	1038	4048	970	733	793
16. 1965-66	557	912	3803	909	606	958
17. 1966-67	453	963	3309	851	406	850
18. 1967-68	742	1175	3802	1096	729	779
19. 1968-69	646	1162	4202	815	688	891
20. 1969-70	779	1174	4408	988	785	792
21. 1970-71	811	1302	4064	1031	743	1191
22. 1971-72	800	1249	3874	1036	788	570
23. 1972-73	748	1225	4337	1009	761	915
24. 1973-74	863	978	4127	920	534	734
25. 1974-75	778	1166	4122	1109	587	593
26. 1975-76	929	1357	4051	1143	724	727
27 1976-77	922	1350	4479	1086	821	829
28. 1977-78*	1065	1429	4775	1126	734	752

*Provisional.

Source : Estimates of Area, Production and Yield of Principal Crops in India, Directorate of Economics and Statistics, Ministry of Agriculture and Irrigation, Government of India, New Delhi.

APPENDIX 4

A NOTE ON THE CONSTRUCTION OF THE REVISED WEIGHTED INDICES OF WHOLESALE AGRICULTURAL PRICES IN U.P.

1. *Objective* : The need for a revised series has arisen in order to bring it in conformity with the rural consumer price index (July 1957 to June, 1958=100) in connection with the preparation of revised series of agricultural parity indices.

2. *Source of Data* : These new indices, like the old series, are based on Friday wholesale price prevalent at district headquarters reported each week by the field staff of this Department posted in each district and also on wholesale prices quoted by selected wholesellers in certain important mandis situated outside district headquarters.

3. *Scope* (*Commodities and Centres*) : The present index is also based on 24 commodities. For each one of these, prices from a number of centres, as given in the enclosed Annexure, are taken into account. The centres selected include all important producing and assembling "mandis" of the State.

4. *Base period* : The agricultural year 1957-58 (July 1957 to June 1958) has been selected as the base period.

5. *Weights* : As in the old series, weights have been used at two stages :

(*a*) Centre-wise weights for working out indices for the State in respect of each commodity.

(*b*) Commodity-wise weights for working out the general weighted index for the State.

5. (*a*) Weight for a centre in case of each commodity is based on the money value of its production in the region represented by that centre ; the total money values have been arrived at by multiplying the quantity of total annual production of the commodity in the region during the agricultural year 1957-58 by the average price of the commodity prevalent at the centre during the same period. Figures of production have been taken from the "Season and Crop Report" 1957-58, brought out by the State Board of Revenue. For the hill portions of the State,

conventional figures of yield, where available, have been utilised. In case of hemp, for which production figures for 1957-58 are not available, figures of production have been arrived at by multiplying the acreage figure for 1957-58 in a region by the figure of normal yield and crop condition. For the five commodities, *viz.*, bhusa, cotton-seed, wheat bran, arhar chuni and mustard cake utilized for cattle-fed index production has been worked out by using information of allied crops given in the "Season and Crop Report, 1957-58". Thus, the yield of bhusa has been estimated at 1.75 times the total production of wheat and barley, of cotton seed at 1.9 times the yield of cotton ginned ; of wheat bran at 2.0 per cent of the yield of wheat, of arhar chuni at 30 per cent of the yield of arhar and of mustard cake at 66 per cent of the yield of mustard and rapeseed. Production figures of ghee in U.P. for 1957-58 are the same as were estimated and utilised in connection with estimates of State Income of U.P. (vide page 1108 of the November, 1955 issue of the Monthly Bulletin of Statistics) ; for determining centre-wise weights for ghee, the estimated value of production for 1957-58 has been distributed over the various centres in the same proportion as in 1938, for which latter year figures were available from the Ghee Development Officer, U.P.

5. (*b*) Weights for a commodity—The total money value of production of a commodity in the State as a whole provides the weight for that commodity. The money value has been calculated on the same lines as indicated under para 5(*a*) above. The total money value of mixed crops, *viz.*, mustard and rapeseed, linseed and til has been allocated to individual centres in the same ratio as for the pure crop. The yield of mixed crops for individual centres has been estimated on the value ratio of pure and mixed crops and taking into account the production figure of the pure-crop under that centre.

6. *Method*: The price index of a commodity is the weighted arithmetic average of its price relatives for the various centres selected for that commodity. The general weighted index as also the price indices of various groups of commodities are the weighted arithmetic averages of the price indices of

selected commodities. Thus, the formula in the case of a commodity index is :—

$$Ic=\frac{\Sigma\, qp}{\Sigma\, qp}\times 100$$

but in practice the computational procedure adopted is a variation of the above, *viz.*,

$$Ic=\frac{\Sigma\left(pq\,\frac{p}{p}\times 100\right)}{\Sigma pq}$$

and the formula in the case of the general weighted index is :

$$\text{General Weighted Index}=\frac{\Sigma\,(pq)\,Ic}{\Sigma\,(\Sigma\, qp)}$$

where *Ic* represents the index of a commodity for the whole State, *q* represents the quantity of production of a commodity in the region in the agricultural year 1957-58 and P its prices at different centres in the current and base periods respectively.

Source : Reproduced from monthly bulletin of Statistics, February 1965, Department of Economics and Statistics, Uttar Pradesh.

APPENDIX 5.1

(*i*) Procedure for obtaining *Reduced Form* of the Distributed Lag Adjustment Model of *Marce Nelove.*

(1) $\overset{*}{X}_t = a+bp_{t-1}+C\text{Y}_{t-1}+gZ_{t-1}+hW_t+U_t$

(2) $X_t-X_{t-1}=B\,(\overset{*}{X}_t-X_{t-1})$

Equation (2) can be written as

$$B\overset{*}{X}_t=X_t-X_{t-1}+BX_{t-1}$$

$$=X_t-(1-B)\,X_{t-1}$$

$$\overset{*}{X}_t=\frac{i}{B}\,X_t-\frac{(1-B)}{B}\,X_{t-1}$$

substituting the value of $\overset{*}{X}_t=\frac{1}{B}\;X_t-\frac{1-B}{B}\;X_{t-1}$

in equation (1)

is $\frac{1}{B} X_t - \left(\frac{1-B}{B}\right) X_{t-1} = a + b\, P_{t-1} + CY_{t-1} + gZ_{t-1} + hW_t + ut$

$$X_t = aB + bBP_{t-1} + Bc\, Y_{t-1} + B_g\, Z_{t-1} + BhW_t + Bu_t + (1-B)\, X_{t-1}$$

$$X = a_0 + b_2\, P_{t-1} + b_3\, Y_{t-1} + b_4\, Z_{t-1} + b_5\, W_t + b_6\, X_{t-1} + V_t$$

where $a_0 = aB$, $b2 = bB$, $b3 = Bc$, $b4 = Bg$, $Bh = b5$, $Bu_t = V_t$ $1 - B = b_6$

(*ii*) *Koyck's Model*

Referring to equation (4) on page 118

Equation 4 $X_t = a_0 + b_2\, P_{t-1} + b_6\, X_{t-1}$

It may be of some interest to note that the above equation 4 could also be derived on the basis of an "adjustment lag" system developed by Koyck*

Let Equation 5. $X_t = a + b_0\, P_{t-1} + b_2\, d\, P_{t-2} + b_0\, d^2\, P_{t-3}$
$+ \text{bod}^3\, P_{t-4} + \ldots\ldots\ldots$

where a and b are constant terms and

$$0 < d < 1$$

If we lag $= n$. 5 by one period we have Equation 6.

$X_{t-1} = a + b_0\, P_{t-2} + b_2\, d\, P_{t-3} + d^2\, P_{t-4} + \ldots\ldots$

Then if we multiply 1 n. 6 by d we get

Equation 7. $dX_{t-1} = ad + b_2 d\, P_{t-2} + b_0\, d^2\, P_{t-3} + b_0 d^3\, P_{t-4} + \ldots\ldots$

Then if we subtract $= n$. 7 from $= n$. 5 we get

$X_t - dX_t - 1 = a\,(1-d) + b_0\, P_{t-1}$

Equation 8. $X_t = a\,(1-d) + b_0\, P_{t-1} + dX_{t-1}$

Thus $= n$. 8 is the same as the $= n$. 4

when no shifter variables are used.

---

* Koyck, L.M., Distributed Lags and Investment Analysis, North Holland Publishing Company, Amsterdam, 1954.

Bibliography

I ARTICLES

Bauer, P.T. & Yamey, B.S., A case study of response to price in an underdeveloped country. *The Economic Journal*, LXIX, 1959.

Behrman J.R., The price elasticity of the marketed surplus of a subsistence crop, *Journal of Farm Economics*, Vol. 47, 1966.

Bhatia, S S., Spatial variations, changes and trends in agricultural efficiency in Uttar Pradesh 1962-63, *Indian Journal of Agricultural Economics*, Jan.-March, 1967, p. 66.

Dantwala, M.L., Incentives and disincentives in Indian agriculture, *Indian Journal of Agricultural Economics*, Vol. 22, 1967.

Devi, P. Kamla., Response of acreage to change in price—A study of Madras State. *The Economic Weekly*, September 19, 1964.

Falcon, W.P., Farmers response to price in subsistence country, *American Economic Review*, Vol. 56, 1964.

Giri, R., Changes in Land use pattern in Madras State, *Agricultural situation* in India, March, 1968.

Griliches, Zvi., Distributed lags—A survey, *Econometrica* Vol. 35, No. I, January, 1967, pp. 16-49.

Jakhade, V.M. & Majumdar, N.A., Response of agriculture producers to price—A case of jute and rice in India, *Indian Journal of Agricultural Economics*, Vol. 19, 1964,

John, P.V., Responsiveness of relative area—Output of sugarcane and rice to changes in their relative prices in Uttar Pradesh, 1954-63. *Indian Journal of Agricultural Economics*, Jan.-March, 1965, pp. 40-47.

Koul, J.L., A study of supply response to price of Punjab crops, *Indian Journal of Economics*, Vol. 47, July, 1967.

Khusro, A.M., The pricing of food in India, *Quarterly Journal of Economics*, Vol. 81, 1967, pp. 271-294.

Klien, L.R., The estimation of distributed lags, *Econometrica*, Vol. 26,1958.

Krishna, Jai and Rao, M.S., Dynamics of acreage allocation for wheat in Uttar Pradesh, *Indian Journal of Agricultural Economics*. Vol. 22, 1967.

Krishna, Raj, Farm supply response in India-Pakistan—A case study of The Punjab Region, The *Economic Journal*, Vol. 73, 1963.

Lipton, M., Should reasonable farmers respond to price changes, *Modern Asian Studies*, Jan. 1967.

Madalgi, S.S., Prices and productivity trend in Indian Agriculture, *Indian Journal of Agricultural Economics*, Vol. 9, 1954

Madan, B.K., Presidential address, Proceedings of 18th conference of the Indian Society of Agricultural Economics, *Indian Journal of Agricultural Economics*, Vol. 13, 1958.

Mollor, J.W., Towards a theory of *Agricultural Development and Economic Growth*, ed. H.M. Southworth and B.F. Johnston (Cornell University Press, Ithaca, New York), 1967.

Nerlove, Marc., The Analysis of changes in Agricultural supply, *Journal of Farm Economics*. Vol. 40, 1958.

Nerlove, Marc., Distributed lags and Estimation of long run supply and demand elasticities, *Journal of Farm Economics*, Vol. 40, 1958.

Pillai, P.P., Impact of Price Movements on Area, Productivity and Production of Paddy in Kerala State, *Agricultural situation in India*, Dec. 1969, pp. 819-822.

Parthasarthy, G., Substitution between sugarcane and paddy in Madras State, *Indian Journal of Agricultural Economics*, Vol. 14, 1959.

Pandey, R K., The Analysis of demand for food-grains, *Indian Journal of Agricultural Economics*, April-June, 1973.

Panse, V.G. Trends in Area and Yield of Principal Crops in India, *Agricultural situation in India*, Vol. VII, No. 3.

Ramamurthy, N., Work Done on Growth Rates in Agriculture in India-A Review, *Agricultural Situation in India*, August, 1967.

Rao, Hanumantha, C.H., Agricultural Growth and Stagnation in India, *The Economic weekly*, February 27, 1965.

Rao, M.S. & Jai Krishna., Price Expectation And Acreage Response For wheat in Uttar Pradesh, *Indian Journal of Agricultural Economics*, Jan.-March, 1965, pp. 20-25.

Rao, M.S. & Jai Krishna, Dynamics of acreage allocation for wheat in Uttar Pradesh—A study in supply response ; *Indian Journal of Agricultural Economics*, Jan.-March, 1967, Vol. 22.

Rao, Subba, Farm Supply Response—A case study of sugarcane in Andhra Pradesh, *Indian Journal of Agricultural Economics*, Vol. XXIV, Jan.-March, 1969, No. I.

Ram, G.S., Supply Response of Cereals in Different States of India, *Agricultural Situation in India*, Vol. XXVIII, Otc. 73, No. 7, p. 469.

Singh, Daroga & Rao, P.R., Estimation of agricultural acreage response relationship : Some methodological Issues, *Indian Journal of Agricultural Economics*, Jan.-March, 1974, Vol. XXIX, No. I.

Sayigh, Y.A., The place of agriculture in economic development, *Agricultural Situation in India*, Annual Number, 1959, p. 445.

Sharma, T.C. & Countinho, O., A Spatio-temporal analysis of wheat crop in Uttar Pradesh, 1960-61 to 1974-75, *Agricultural Situation in India*, May, 1978.

Sethi, J.D., Some analytical and quantitative aspects of agricultural supply and substitution function in India, *Indian Journal of Agricultural Economics*, April-June, 1966, Pp. 16-30.

Sivaraman, B , The strategy of food production ; *Agricultural Situation in India*, October, 1967.

Singh, I.J. & Pranesh Kumar, Impact of price and price variability on acreage allocation in Haryana, *Indian Journal of Agricultural Economics*, Vol. XXXI, No. 2, April-June, 1976.

Venkataramanan (L.S.) & Prahaladachr, (M) ; Study of cropping pattern changes in Andhra Pradesh, during 1950-75. *Indian Economic Review*, Oct., 1978, No. 2.

Warriner, Doreen, Land reform and economic development, in *Agriculture in Economic Development*, ed. Carl Eicher, Witt, L. pp. 272.

BOOKS

Behrman, J.R., *Supply response in Underdeveloped agriculture*, (Contributions to economic analysis LV) North-Holland, Amsterdam, 1968.

Bansil, P.C , *Agricultural problems of India*, Vikas Publishing House, Pvt. Limited, New Delhi.

Christ, C.F., *Econometric models and methods*, Wiley & Sons, New York, 1966.

Coale and Hoover., *Population growth and economic development in low income countries : A case study of India's* prospects pp. 120-139.

Chaing, *Economic Dynamics and Integral Calculus.*

Croxton & Cowden. Applied General Statistics.

Das Gupta, S., *Agriculture Producer's Rationality and Technical Change*, Asia Publishing House, 1970.

Daya Krishna ; *New Strategy*, New Heights Publishers and Distributors, 1971.

Dean, E., *The Supply Responses of African Farmers*, North Holland, Amsterdam, 1966.

Goldberger, A.S., *Econometric Theory*, Wiley & Sons, New York, 1964.

Heady, E.G., *Economics of Agricultural Production and Resource Use* Prentice Hall Inc, New Jersey, 1957, pp. 475-496.

Johnston, J., *Econometric Methods* McGraw Hill Book Company, New York, 1963.

Kendall, M.G. & A. Stuart, *Advanced Theory of Statistics*, Vol. I, Charles Griffin, London, 1961.

Koyck, L.M., *Distributed Lags and Investment Analysis*, North Holland, Amsterdam, 1954.

Kmenta, Jan ; Elements of Econometrics.

Mellor, J.W. *The Economics of Agricultural Development*, Vora & Co. Publishers, Pvt. Ltd., Bombay.

Marc Nerlove., *The Dynamics of Supply* ; Estimation of *Farmers' Response to Price*, Baltimore, The Johns Hopkins Press, 1958.

National Council of Applied Economic Research, New Delhi, Structure and behaviour of Prices of Food grains.

Techno-economic Survey of Uttar Pradesh 1965.

Narain, Dharm, *Impact of Price Movements on Acreage under selected crops in India,* Cambridge University Press, London.

Rao, C.R. *Advanced Statistical Methods in Biometric Research,* Wiley and Sons, New York, 1952.

Schultz, T.W. *Economic Crises in World Agriculture,* The University of Michigan Press, Ann Arbor, Michigan, 1965.

Transforming Traditional Agriculture, Yale University Press, London, 1964.

Stigler, G., *Theory of Price,* Macmillan, New York, 1947.

Tinbergen J. *Econometrics,* Blakiston, New York, 1951.

Wilks, S.S. *Mathematical Statistics,* Wiley and Sons, New York, 1962.

Government Publications

(*i*) *Government of U.P.*

1. Season and Crop Reports,
2. Statistical Abstracts
3. Economic Surveys,
4. Series of Wholesale and Farm Harvest Prices.

(*ii*) *Government of Madhya Pradesh*

Agricultural Development in M.P. 1950-51 to 1975-76, Part I & Part II., Agro-Economic Research Centre, Jabalpur, M.P.

(*iii*) *Government of India Publications*

Estimates of Area, Production & Yield of Principal Crops in India, D.E.S. Minister of Agriculture, New Delhi.

National Income Estimates, C.S.O. New Delhi.

Reports of National Commission on Agriculture, Ministry of Agriculture & Irrigation, New Delhi.

Fertilizer Statistics, Fertilizer Association of India, New Delhi.

Population Projections of Indian States, Registrar General (Cenus) of India. New Delhi.

Bulletins on Food Statistics, Ministry of Agriculture & Irrigation.

Index

Adjustment, estimated coefficient of, 124*t*
Agricultural commodities, prices of, 78
Agricultural development, 21ff, 67 ; programmes of, 48 ; output, 78 ; price analysis, 50 ; prices, 78 ; products, supply of, 103
Agricultural Production Team, 8 ; Sector, 3
Agro-climate differences, 54
America, United States of, 1, 5
Annual crops ; 106, plans, 2
Area, growth pattern of, 21 ff ; percentage change in, 31*t*
Artificial, irrigation facilities, 67
Askari, Hossein, 106

Barley, 17, 25, 38, 48, 91
Baroda, All-India Agricultural Economics Conference in, 1
Bauer, P.T., 6
Behaviour of Farm harvest prices, 81
Behrman, J.R., 106
Bihar, 10
Brazil, 127

Cereals, projections of, 132
Chemical fertilizers, 74 ; distribution of, 74
Cropping intensity, 62
Cropping pattern, 56 ; changes of, 54
Cummings, Johan Thomas, 106

Data, basis of price, 92 ; farm harvest price, 82
Dehra Dun, 10

Economics and statistics, Directorate of, 18

Farm commodities, 95
Farm harvest prices, movements of, 81, 88 ; selected crops of, 95 ; land resources, 67 ; price indices of, 84*t*, 89*t* ; sector, 4
Five Year Plans, 1, 2, 6
Food availability, projections of, 127
France, 1

Galbraith, J.K., 104
Gram, 17, 25, 38, 49
Green Revolution, 8, 74
Growth rates of area, 51 ff ; at a glance, 94

Growth of yield trends, 38
Gupta, S.C., 106

Haryana, 10
Higgins, Benjamin, 5
Highest price index, 93
High yielding varieties programme, 61, 74 ; crop varieties, 55
Himachal Pradesh, 1
Himalayas, 10

Independent variable, gross irrigated area, 110 ; logged acreage and relative, 110 ; rainfall, 110 ; relative lagged area, 110
India, 1, 7, 8
Investigation, area selected for, 9 ff ; objectives of, 13 ff
Irrigated area under selected crops, 71 ; by source of irrigation, 72*t*
Irrigation potential, growth of, 67

Japan, 1, 6
John, P.V., 6, 106

Keynes, J.M., 79
Krishna, Raj, 79, 105, 106, 107
Kovck's model, 147
Kumaon Division, 10
Kumar Chandresh, 106

Lag adjustment model, 145
Land use categories, 62
Lewis, W. Arthur, 5
Linear regression equation for wheat price, 86
Lipton, Michael, 80

Madhya Pradesh, 10
Maize, 17, 25, 38 ; coefficient of variation of price series of, 93
Majid, A., 106
Mason, Edward, 80
Ministry of Agriculture and Irrigation, 18

The Model, Format of, 111ff

Myrdal, Gunnar, 5

Nanda Devi, 10
Narain, Dharm, 81, 106
Nepal, 10
Nerlove, Marc, 104
Nerlovian adjustment, models, 108 ; distribution lag model, 104, 105

Parthasarthy, G., 106
Physical features of Uttar Pradesh, 10
Post harvest prices, index of, 107
Price, whole sale 82 ; behaviour analysis of, 101, data, 82 ; indices, 99 ; mechanism, 103 ; movements, 95 ; policy, 79

Rajasthan, 10
Random Crop Cutting Surveys, 20
Rao, V K.R.V., 78
Revised weighted indices, construction of, 143
Rice, 17, 21, 33, 71 ; acreage response equation of, 112, 119
Role of Price mechanism, 78

Satyanarayan, Y., 106
Schultz, T.W., 4, 80
Season and crop report 1957-58, 144
Selected crops, 39, 103 ; production and yield of, 21 ; three yearly moving averages of volume output of, 36*t* ; variability of yields of, 51*t*

Selected farm crops, 66; commodities, movements of wholesale prices of, 95
Sen, S.R., 1
Smith Adam, 4
Soviet Socialist Republics, Union of, 1
Sown area, per capita, 65
State economy, importance of agriculture, 12; educational achievement of the, 13
Sugarcane, 12, 17, 24, 33: acreage response equations of, 122

Tibet, 10
Total irrigated area, 71
Trends of area production and yield, study of, 17 ff

Unculturable weste land, 62
United Kingdom, 1, 5
United Nations, 3
Uttar Pradesh, 9, 10, 55, 71, 82, 106; annual area, variation of selected agricultural crops, 22*t*; annual rates of growth of out-put of selected agricultural commodities in. 34*t*; area under selected agricultural crop in, 138; area under HYVP crops in, 76*t*; changes in cropping pattern in, 56, 61; cropping pattern in, 57; distribution of monthwise rainfall in, 68*t*; distribution of fertilizer in, 75*t*; growth of population in, 118; growth rates of area production and yield of various groups in, 52*t*; Land use pattern in, 61, 63*t*; literacy in, 13*t*; net availability of food-grains in, 130; percapita area sown in, 65ff; per hectare yield of selected agricultural crops in, 142; produuction of food-grains in, 129; production of selected agricultural crops in, 40*t*. 140; sowing and harvesting seasons of selected agricultural crops in, 66*t*; yield of selected agricultural crops in. 44*t*; yearwise dispersion of rainfall in, 70

Wealth of Nations, 4
Weather and crop calendar, 66
Wheat, 17, 24, 30, 60, 71; acreage response equations of, 120; price indices of, 96
World Agriculture, economic crises in, 4

Yamuna river, 10
Yamuna-gangetic plain, 10
Yield of selected crops, variability in, 50 ff